Control in Bioprocessing

Control in Bioprocessing

Modeling, Estimation and the Use of Soft Sensors

Pablo Antonio López Pérez
Escuela Superior de Apan, Universidad Autónoma del Estado de Hidalgo, México

Ricardo Aguilar López
Department of Biotechnology and Bioengineering, Center for Research and Advanced Studies (Cinvestav), México

Ricardo Femat
Department of Applied Mathematics, Institute for Scientific and Technological Research of San Luis Potosi (IPICYT), México

Registered Offices
John Wiley & Sons, Inc., 111 River Street, Hoboken, NJ 07030, USA
John Wiley & Sons Ltd, The Atrium, Southern Gate, Chichester, West Sussex, PO19 8SQ, UK

Editorial Office
The Atrium, Southern Gate, Chichester, West Sussex, PO19 8SQ, UK

For details of our global editorial offices, customer services, and more information about Wiley products visit us at www.wiley.com.

Wiley also publishes its books in a variety of electronic formats and by print-on-demand. Some content that appears in standard print versions of this book may not be available in other formats.

Library of Congress Cataloging-in-Publication Data Applied for

HB ISBN: 9781119295990

Cover Design: Wiley
Cover Image: © Reptile8488/Getty Images

Set in 9.5/12.5pt STIXTwoText by SPi Global, Chennai, India

Printed and bound by CPI Group (UK) Ltd, Croydon, CR0 4YY

10 9 8 7 6 5 4 3 2 1

Contents

Preface

The purpose of this book is to present the different approaches most commonly employed in the control of bioprocesses. It aims to develop in some detail the bases and concepts of bioprocesses related to the control theory introduced in basic principles of mathematical modeling in bioprocesses. From this viewpoint, the systems approach to bioengineering and bioprocessing, with its current focus on the development of mathematical models and their analysis, is a logical sequel that the control theory that will play a relevant role in understanding the mechanisms of cellular and metabolic processes. It concerns specifically applications in modeling, estimation, and control of bioprocesses. Consequently, this book presents key results in various fields, including: dynamic modeling, dynamic properties of bioprocess models, software sensors designed for the online estimation of parameters and state variables, and control and supervision of bioprocesses. The book is divided into three sections.

Part I: Overview of the Control and Monitoring of Bioprocesses and Mathematical Preliminaries, contains Chapters 1 and 2. Chapter 1 is a general overview of the control and monitoring of biotechnological processes. Chapter 2 introduces the mathematical framework necessary for the analysis and characterization of bioprocess dynamics. In other words, Chapter 2 deals with the mathematical approach we follow to describe the evolution in time of the bioprocess under consideration. Therefore, understanding and formalizing the role of nonlinearity is indeed one of the greatest challenges in the study of living systems, mainly in bioprocesses. In engineering practice two of the most important sources of modeling error are the presence of nonlinearities in the system and a lack of exact knowledge of some of the system parameters, therefore, it is necessary to describe the properties of the state of the dynamics of the nominal model under the theory of systems.

Part II: Observability and control concepts, contains Chapter 3 on state estimation and observers and Chapter 4 focusses on control of bioprocess. Chapter 3 introduces the reader to the observability concepts that are the basis to designing online estimation algorithms (software sensor) for bioprocess. The observability conditions for bioprocess models from local linearization, differential geometric and algebraic differential approaches are established. Chapter 4 reviews the controllability concepts that are the basis for designing automatic feedback control schemes for bioprocesses. A clear explanation is developed from the classical linear schemes to advanced robust algorithms with application in bioprocess systems. However, performance may degrade when they are applied to highly nonlinear processes, which are the fact rather than the exception in the chemical and biochemical process industry.

Part III: Software sensors and observer-based control schemes for bioprocess. This last part deals with application cases in Chapters 5–10. Chapter 5 covers the dynamical behavior of a three-dimensional continuous bioreactor. The dynamic behavior of a two-dimensional model of a continuous bioreactor was studied in this chapter. The objective of the analysis under the control of feedback allows the most suitable regions to be found and to model where the best performance of the bioreactor operates. Chapter 6 reviews observability analysis applied to 2D and 3D bioreactors with inhibitory and non-inhibitory kinetics models. The results indicate that the proposed model can be applied for simulation in different conditions of operation for possible instrumentation, estimation and control from laboratory scale up to semi-pilot scale. Chapter 7 introduces the production system myco-diesel for implementation of "quality" of the observability. Myco-diesel is a new alternative and, compared to traditional fossil fuels, an environmentally sustainable biofuel source due to reduced gas emission. It is made from renewable bioprocess sources such as vegetable oils, animal fats, and microorganism culture. Chapter 8 is about the regulation of continuously stirred bioreactors via modeling error compensation. The aim of this chapter is to control the nonlinear behavior of a class of continuously stirred bioreactors with regulation purposes. From the above, a linearized representation of the state space bioreactor's model is obtained via standard identification processes employing a step disturbance in the control input. Chapter 9 reviews the development of virtual sensors based on the just-in-time model for monitoring of biological control systems. Real-time monitoring of physiological characteristics during a cultivation process is of great importance in bioprocesses. Biological control involves the use of beneficial organisms for metabolite production that reduces the negative effects of plant pathogens disease suppression. Finally, chapter 10 discusses virtual sensor design for state estimation in a photocatalytic bioreactor for hydrogen production. This chapter is focused on the design of a virtual sensor for a class of continuous bioreactors to estimate the production of hydrogen. The proposed mathematical model is suitable for predicting concentrations of biomass, acetate, cadmium in liquid, sulfate, lactate, carbon dioxide, sulfide, and cadmium sulfide, as well as hydrogen production.

Bioprocess modeling and control still offers interesting perspectives to obtain automatization solutions for the aerobic and anaerobic bioprocesses.

Part I

Overview of the Control and Monitoring of Bioprocesses and Mathematical Preliminaries

1

Introduction

This chapter aims to develop in some detail the basis and concepts of the bioprocesses related to the control theory introduced in the basic principles of mathematical modeling in bioprocesses. From this viewpoint, the systems approach bioengineering and bioprocessing, with a current focus on the development of mathematical models and their analysis, which logically results in the control theory playing a relevant role in the understanding of the mechanisms of cellular and metabolic processes. It concerns specifically applications in modeling, estimation, and control of bioprocesses. Consequently, there are many exciting opportunities for control experts who want to shift their interest to bioprocesses.

1.1 Overview of the Control and Monitoring of Bioprocesses

1.1.1 Why Nonlinear Control in Bioprocesses?

Nowadays, many proteins and metabolites are produced by genetically modified microorganisms applied to bioprocesses. Since the physical, biochemical, and genetic properties of the bacterium *Escherichia coli* are the best known, it is the most generic host microorganism used for the production of bio-products.

The success of this process is ensured by appropriate control of the feeding rate. In other words, under-feeding causes productivity loss and starvation whereas over-feeding leads to carbon nutrient accumulation or by-product formation, such as acetate [1, 2].

Current bioprocess control technology and opportunities include biotherapeutics, specialty chemicals, and reagents such as diagnostics, biochemicals for research, and enzymes for the food and consumer markets.

Products and services that depend on bioprocessing can be grouped broadly into antibiotics, therapeutic proteins, polysaccharides, vaccines and diagnostics, value-added food, and agricultural products, as well as fuels, specialty products and industrial chemicals, and fibers from renewable resources (see Figure 1.1) [13].

Of course, any bioprocess may be considered as a system, being built up of large numbers of individual organisms or cells, each of which may show a far simpler behavior than the whole system or may show a far more complex one.

Despite the diversity of organisms, they all possess the following specific features that must be taken into account in studying and designing the bioprocesses: cells self-organize,

Control in Bioprocessing: Modeling, Estimation and the Use of Soft Sensors, First Edition.
Pablo Antonio López Pérez, Ricardo Aguilar López, and Ricardo Femat.
© 2020 John Wiley & Sons Ltd. Published 2020 by John Wiley & Sons Ltd.

The Importance of Control Theory in Bio-process

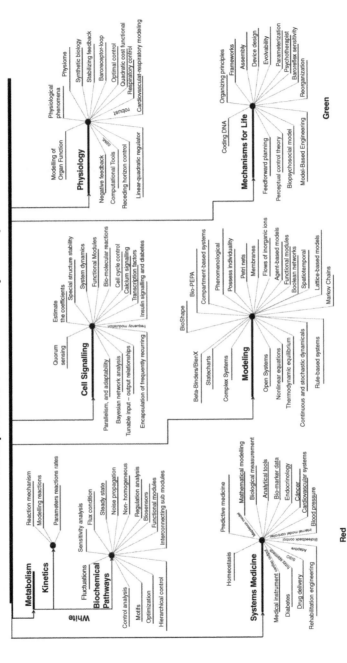

Figure 1.1 Opportunities for control by topics [3–12].

interaction between intra-cellular processes and inter/extra-cellular systems, metabolic processes, signaling, and regulatory bioprocesses at different molecular levels, organized and regulated interplay between different pathways or network modules, structured and non-structured models (metabolic or genetic level), inorganic, and organic matter for the biosynthesis of biological macromolecules, homogeneous and heterogeneous systems, flows of matter and energy, for example, of inorganic ions and organic molecules. An outline of these topics is shown in Figure 1.1.

Achievements in modern biology have revealed numerous facts related to the structure and regulation of many intracellular systems. Schemes of bioprocesses are complex, they have nonlinear kinetics, a chemical structure and, in most cases, the molecular structures of the components of processes are examined, including the bio-regulators. This makes it possible to construct mathematical computer models that allow formalization of the knowledge of complex biological objects. The degree of specification of models can be different depending on the goal of modeling and on the degree of completeness of the examination of the objects. If the modeling is aimed at control, for example, an efficiency increase in the output of a biotechnological process is desired, and then it is often sufficient to consider individual blocks as components and examine stationary states of a system. They are modeled by "constructors", that is, programs that automatically write differential equations according to a prescribed scheme of processes and expressions for the rates of individual reactions. In investigating such complex systems, the theory of metabolic control has a deserved good reputation. If an object is thoroughly examined, mathematical models become an effective method of fundamental research. By solving inverse problems, they allow the estimation of kinetic and physical parameters of a holistic system, which is impossible in experimentation without fractionating a system. In complex biological systems, the latter leads to the modification of the functional activity [14].

Control of bioprocesses is a complicated task for, at least, the following three reasons:

i. The process' complexity, nonlinearity and non-stationarity which make modeling and parameter estimation particularly difficult.
ii. The scarcity of on-line measurements of the component concentrations (essential substrates, biomass, and products of interest) [14].
iii. Furthermore, almost all such bioprocesses are nonlinear systems. Because of the large number of interactions in such systems, some are bound to be nonlinear. Given the large number of species, strains, and consortia, they are almost certain to show chaotic behaviors under a wide range of conditions. Complex systems may show roughly four types of behavior:
   - steady state,
   - periodic,
   - complex, and
   - chaotic.
   steady state is the simplest; the system is unmoving into a particular state. Though there may be some initial oscillations, these die out and the system settles down into its final state [15].

A dynamical system is a simple system that can be characterized by

i. a set of parameters the values of which define its state at a given point in time, and
ii. a set of mathematically specified rules defining the change of state of the system with time.

The rules of dynamical systems can be specified as differential equations, defining the rate of change of each of the parameters describing the system, as a function of the current state of the system. This definition is broad indeed, and many systems in biotechnology, bioengineering, medicine, bio-economics, and the social sciences may be described and studied as dynamical systems. The extracellular and intracellular metabolite concentrations from industrial bacteria, either strain [16, 17], mixed cultures (consortia) [18], or ecosystems [19] can be described in terms of dynamical systems. The key feature of the dynamics of these systems is that they show autocatalytic or inhibitory loops: the presence of a biomass is needed to make more of that kind of biomass.

Furthermore:

i. some species may inhibit other species by secretion of toxins by competition;
ii. some species might also enhance growth by producing metabolites that serve as food for other species by cooperation;
iii. or may remove or reduce concentrations of toxic substances by symbiosis.

In systems that are far from the thermodynamical equilibrium, such autocatalytic and inhibitory loops produce just the type of nonlinear dynamics that can produce highly complicated and chaotic behaviors [15].

In practice, all ecosystems are far from thermodynamic equilibrium, since large fluxes of energy or food pass through them; only death (a point attractor of any ecosystem) corresponds to thermodynamic equilibrium.

For these reasons it may be assumed that techniques for analysis and modeling of nonlinear dynamical systems are, in general, appropriate tools for the study of bioprocesses including bacterial, fungal, mammalian cells, or insect or plant cells. The following examples are developed (modeling and simulation) to represent the diversity and complexity of some bioprocesses:

Example 1.1 *A Mathematical Model for Cadmium Removal Via Sulfate-Reducing Process*

Sulfate reducing bacteria (SRB) have been used to solve a number of environmental problems, e.g. removal of metals from wastewater through the production of biogenic sulfides, followed by metal precipitation. A mathematical model was developed to describe the kinetics of cadmium (Cd^{2+}) removal in a batch bioreactor. The classical growth with reduction of sulfate to H_2S is described by Eqs. (1.1–1.6). The Levenspiel inhibition model was modified to describe the reduction of sulfate. The model considers the inhibitory effect of sulfide (H_2S) on microbial growth [20–22]. The mass balance of the various concentrations gives the following set of equations:

Sulfate (S) mass balance:

$$\frac{dS}{dt} = -\frac{k_{spx}}{Y}\left(1 - \frac{P}{k_p}\right)^{\alpha}\left[\frac{S}{k_s + S}\right]XL^{\varepsilon} \tag{1.1}$$

Sulfide (P) mass balance:

$$\frac{dP}{dt} = \frac{k_{spx}}{Y_p}\left(1 - \frac{P}{k_p}\right)^{\alpha}\left[\frac{S}{k_s + S}\right]XL^{\varepsilon} \tag{1.2}$$

Biomass (X) balance:

$$\frac{dX}{dt} = k_{spx}\left(1 - \frac{P}{k_p}\right)^\alpha \left[\frac{S}{k_s + S}\right] \left[\frac{Cdl^{2+}}{k_{Cd} + Cdl^{2+}}\right]^\eta X L^\varepsilon - k_d X L^\varepsilon \tag{1.3}$$

Lactate (L) mass balance:

$$\frac{dL}{dt} = -\frac{k_{LA}}{Y_{L/X}} \left[\frac{k_{ace}}{A + k_{ace}}\right] \left[\frac{L^\delta}{k_{lac} + L^\delta}\right] X \tag{1.4}$$

Acetate (A) mass balance:

$$\frac{dA}{dt} = \frac{k_{LA}}{Y_{A/X}} \left[\frac{k_{ace}}{A + k_{ace}}\right] \left[\frac{L^\delta}{k_{lac} + L^\delta}\right] X \tag{1.5}$$

The mass balance describing the removal of Cd^{2+}:

$$\frac{dCdl}{dt} = -k_{u\ max\ Cdl}\left(1 - \frac{Cdl}{k_p}\right)^\gamma \left[\frac{CdS}{k_1 + CdS + \frac{CdS^2}{k_2}}\right] P \tag{1.6}$$

In addition, the model showed that the H_2S production rate and initial concentration of Cd^{2+} are key operating variables in a bioreactor. *Desulfovibrio alaskensis* 6SR was able to remove more than 99.9% of cadmium in a batch process (see Figures 1.2 and 1.3), where the initial concentration was $170\ mg\ l^{-1}$ (see Figure 1.4).

Figure 1.2 Comparison of: (a) experimental sulfate, (b) experimental biomass, and (c) experimental sulfide with predictions of the model (−).

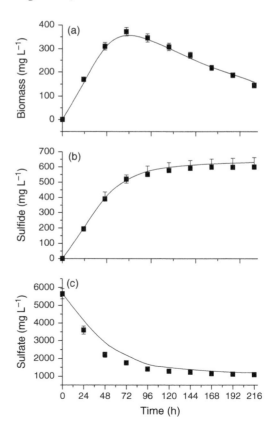

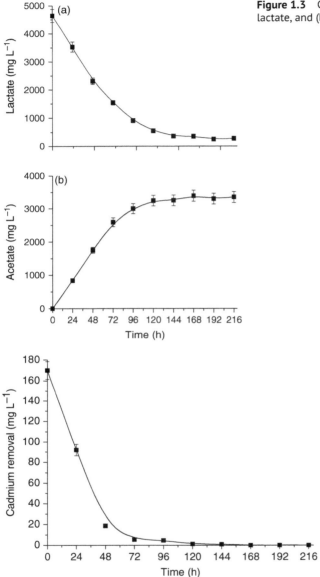

Figure 1.3 Comparison of: (a) experimental lactate, and (b) experimental acetate.

Figure 1.4 Comparison of experimental Cd^{2+} with predictions of the model (−).

Example 1.2 *Simple Metabolic Model*

As an example, we consider a simple metabolic model, illustrated by the Figure 1.5 [23]. Both types of feedback loop are important in biological systems, and both can produce chaos, the mathematical complexity of which often produces strange, beautiful, and totally unexpected patterns that have only begun to be explored using the computational capabilities of modem electronic computers. Examples of such systems are found in catabolic pathways in which the first step involves coupling of adenosine triphosphate (ATP) hydrolysis to activation of a substrate [24, 25].

Figure 1.5 Simple metabolic model with feedback.

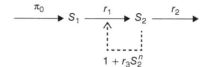

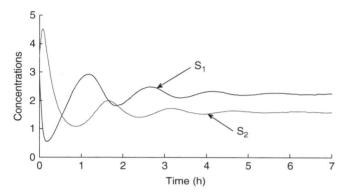

Figure 1.6 Temporal evolution of the variables described by Eqs. (1.7 and 1.8). It is demonstrated, that increasing the nonlinearity in the model leads to the sustained oscillatory behavior.

This scheme is motivated by the "turbocharged" positive feedback aspect of the glycolytic chain in which the ATP output is used to produce more ATP. The potential for oscillations can be inferred from the model structure: the concentration of species S_2 builds up, causing further buildup until the pool of S_1 is depleted. The S_2 level then crashes until more S_1 is available, and so on. Although this intuitive argument indicates the potential for oscillatory behavior, it cannot predict the conditions under which oscillations will occur (see, Figures 1.6–1.9). Positive feedback tends to reinforce the effects of changes in the input signal, producing instability, oscillation, and even catastrophic destruction of the system [26].

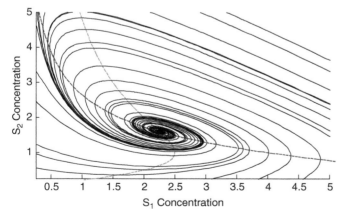

Figure 1.7 The figure shows phase-plane, we see that the trajectories are attracted to a cyclic track, called a limit cycle, the parameter values used are $\pi_0 = 8$, $r_1 = 1$, $r_2 = 5$, $r_3 = 1$, and $n = 2$.

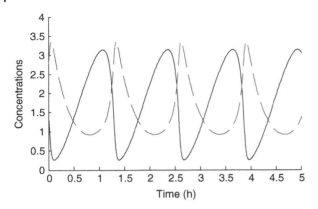

Figure 1.8 Temporal evolution of the variables described by Eqs. (1.7 and 1.8). The parameter values used are $v_0 = 8$, $k_1 = 1$, $k_2 = 5$, $k_3 = 1$, and $q = 2$.

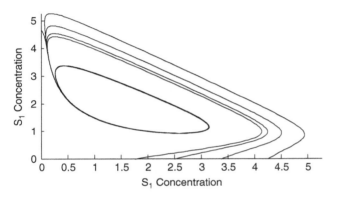

Figure 1.9 The figure shows Phase-plane, we see that the trajectories are attracted to a cyclic track, called a limit cycle, the parameter values used are $v_0 = 8$, $k_1 = 1$, $k_2 = 5$, $k_3 = 1$, and $q = 2$.

The equations describing the system are given by

$$\frac{dS_1}{dt} = \pi_0 - r_1 S_1 (1 + r_3 S_2^n) \tag{1.7}$$

$$\frac{dS_2}{dt} = r_1 \sigma_1 (1 + r_3 S_2^n) - r_1 S_2 \tag{1.8}$$

The qualitative behavior of the system changes as the parameter values shift [23].

Example 1.3 *Glucose Regulatory System*
Diabetes is a dysfunction of the equilibrium in the blood glucose homeostasis. This is caused by an autoimmune attack on beta cells secreted by the pancreas (Type 1) or by the inadequate supply or function of β-cells in counteracting the fluctuations of high and low blood glucose within the body (Type 2) [27, 28]. The experimental data suggests that multiple defects are required for the onset of type 2 diabetes [29]. Glucose dynamics is released into the blood by the liver and kidneys, removed from the interstitial fluid by all the cells of the body, and distributed into many physiological compartments (e.g. arterial blood, venous blood, interstitial fluid). Experimental studies have shown that despite such

a complex distribution, slow glucose dynamics (on a time scale of hours and larger) can be modeled using a one-compartment approximation [30, 31]. Mathematical modeling of diabetes has focused predominantly on the dynamics of a single variable, usually glucose or insulin level [30]. Those models were usually used for measuring either rates (glucose, insulin production/uptake) or sensitivities (insulin sensitivity, glucose effectiveness) [31]. The model introduced by Topp et al. [29] incorporates β-cell mass, insulin, and glucose dynamics, and gives rises to a stiff system of ordinary differential equations (ODEs). The resulting system has two stable sink type equilibrium points and a saddle point. The description of the model presented in this section follows the original paper of Topp et al. [29].

Although simplistic, the systemic reduced-order model allows us to show how it is possible to study a biosystem in the context of this book. The mathematical model studied gives the following set of equations:

$$\frac{dI_p}{dt} = -f_1 - E\left(\frac{I_p}{v_p} - \frac{I_p}{v_i}\right) - \frac{I_p}{t_p} \qquad (1.9)$$

$$\frac{dI_i}{dt} = -E\left(\frac{I_p}{v_p} - \frac{I_p}{v_i}\right) - \frac{I_i}{t_i} \qquad (1.10)$$

$$\frac{dG}{dt} = G_{in} - f_2 - f_3 * f_4 + f_5 \qquad (1.11)$$

$\frac{dI_p}{dt}$: The amount of insulin in the plasma with respect to time.

$f1$: The pancreatic insulin production controlled by the glucose concentration.

E: A constant transfer rate for exchange of insulin between plasma and remote compartments.

Ip: The amount of insulin in the plasma.

Ii: The amount of insulin in the intracellular space.

vp: The distribution volume for insulin in the plasma.

vi: The effective volume of the intracellular space.

tp: The insulin in the plasma as a time constant.

$\frac{dG}{dt}$: The amount of glucose in the body with respect to time.

G_{in}: The influx of glucose in the plasma and intracellular space at an exogenously controlled rate.

f_2: The insulin-glucose independent utilization (uptake by the brain and nerve cells).

f_3: The glucose utilization by the muscle and fat cells.

f_4: The relationship between the plasma insulin concentration and the cellular glucose uptake.

$$\frac{dx_1}{dt} = \frac{3}{t_d}(I_p - x_1) \tag{1.12}$$

$$\frac{dx_2}{dt} = \frac{3}{t_d}(x_1 - x_2) \tag{1.13}$$

$$\frac{dx_3}{dt} = \frac{3}{t_d}(x_2 - x_3) \tag{1.14}$$

$$f_1 = \frac{R_m}{1 + e^{\left(C_1\frac{G}{V_g}/\alpha_1\right)}} \tag{1.15}$$

$$f_2 = U_b\left(1 - e^{\left(-G/(C_2*V_g)\right)}\right) \tag{1.16}$$

$$f_3 = \frac{G}{C_3 * V_g} \tag{1.17}$$

$$f_4 = U_o + \frac{U_m - U_o}{1 + e\left(-\beta ln\left(\frac{I_i}{C_4\left(\frac{1}{V_i} + \frac{1}{Et_i}\right)}\right)\right)} \tag{1.18}$$

$\frac{dx_1}{dt}$: The inhibition of hepatic glucose production via the insulin stimulating pancreatic insulin production.

$\frac{dx_2}{dt}$: Physiological action of insulin on the utilization of glucose correlated with the concentration of insulin in a slowly equilibrating intercellular compartment rather than with the concentration of insulin in the plasma.

$\frac{dx_3}{dt}$: Time lag between the appearance of insulin in the plasma and its inhibitory effect on the hepatic glucose production.

t_d: The response of the hepatic glucose production to changes in the plasma insulin concentration involves a time delay.

x_1, x_2, x_3: Represents the relationship between the time delays of insulin in plasma and its effect on the hepatic glucose production.

R_m: The rate of glucose metabolism within the cell.

C_1: A given parametric value attained by experimental tests that pertains to the process within the function.

G: Glucose.

V_g: The volume of Glucose bα_1: A given parametric value attained by experimental tests that pertains to the process within the function.

U_b C_2: Are given parametric values attained by experimental tests that pertains to the process within the function.

V_g: The Volume of Glucose.

C_3: A given parametric value attained by experimental tests that pertains to the process within the function.

U_o t_i C_4 β $V_i U_m$: Are all given parametric values attained by experimental tests that pertain to the process within the function.

$$f_5 = U_o + \frac{R_g}{1 + e\left(\alpha\left(\frac{x_3}{V_p} - C_5\right)\right)}$$

(1.19)

R_g: A given parametric value that denotes rate of glucose.
α: A given constant transfer rate.
C_5: A given parametric value attained by experimental tests that pertains to the process within the function.

The effect of the frequency of the oscillations on the extent of the inhibition of the hepatic glucose production needs further experimental investigation. Such experiments should be designed so that only the frequency is varied, whilst the mean value of the plasma insulin remains constant (see, Figure 1.10). Furthermore, it may be desirable to replace the variables for the time delay between the insulin in plasma and its effect on the hepatic glucose production by physiologically meaningful state variables.

Such a mechanism could result from effects of insulin on the pancreatic alpha cells, adipocytes, and muscle cells. In adipocytes insulin inhibits lipolysis, thus decreasing the level of glycerol and free fatty acids that reach the liver (see, Figure 1.11). The decreased concentration of glycerol reduces gluconeogenesis, whereas the lower concentration of free fatty acids suppresses the glycogen degradation into glucose [32, 33].

1.2 Improvements to Bioprocesses Productivity

Advances in genetic engineering have, over the past three decades, generated a wealth of novel molecules that have redefined the role of microbes, and plant–microbial consortium systems, in solving pharmaceutical, environmental, industrial, and agricultural problems. While some products have entered the marketplace, the difficulties of doing so and of complying with federal mandates of: safety, purity, potency, and efficacy have shifted the focus from the term genetic to the term engineering [34].

Here are some examples of some of the industrially efficient tools now coming from the application of bioprocesses and biotechnology (Table 1.1):

The bio-production systems allow major improvements in both economic profitability and environmental performance reducing its environmental impact. This transition from the laboratory to production is the basis of bioprocess engineering and involves a careful understanding of the conditions most favored for optimal production, and the duplication of these conditions during scaled- up production [34].

In biotechnology aimed at improving productivity three approaches are considered:

i. Biochemical/microbiological approach
 - selection of microorganisms/nutrients
 - genetic modifications.
ii. Bioprocess engineering approach
 - operating modes/conditions
 - efficient techniques/bioprocesses.
iii. Bioprocess control/instrumentation approach
 - productivity maximization via an on-line optimizing operation of the bioprocess (on the basis of a dynamical model of the process).

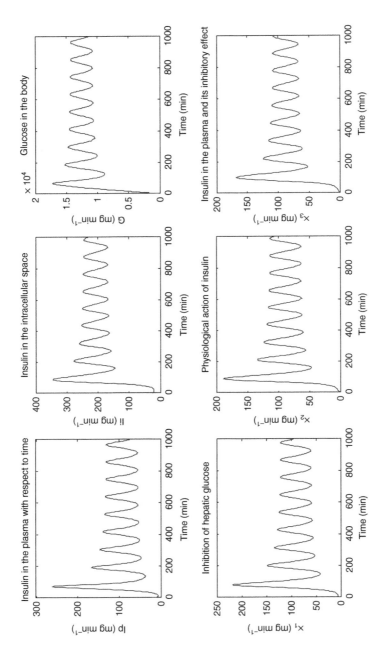

Figure 1.10 Temporal evolution of the variables concentration of the system described by Eqs. (1.9–1.19). Results from the improved model [29]. Plasma glucose and plasma insulin concentrations during a simulated constant glucose infusion with the rate of $216 \, mg \, min^{-1}$. *Source:* Reproduced with permission of Elsevier.

Figure 1.11 The figure shows a phase-plane, we see that the trajectories are attracted to a cyclic track, called a limit cycle, the parameter values used are: $C_1 = 2000\,\text{mgl}^{-1}$, $C_2 = 144\,\text{mgl}^{-1}$, $C_3 = 1000\,\text{mgl}^{-1}$, $C_4 = 80\,\text{mUl}^{-1}$, $C_5 = 26\,\text{mUl}^{-1}$.

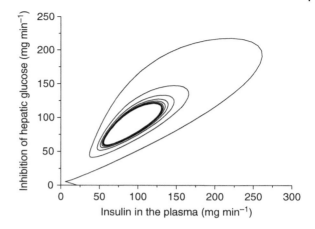

Table 1.1 Bioprocess based sustainable production [34].

Bio-product	Application	Improvement
Extracted enzymes (micro-organisms, plants and animals)	Catalyze chemical reactions	High efficiency and specificity
Microreactor	Metabolic engineering	High yield
Biomass	Chemicals, fuels and materials that are renewable	No net emissions of greenhouse gases
Biopolymers and bioplastics	Plastics, food, medicine, waste water	Reduce environmental impacts.
Biofertilizer	Agriculture sector	Reduce environmental impacts

The aim of bioprocesses is to apply and improve natural systems by genetical manipulation of cells and their environment to produce them industrially and of high quality [35].
Systems used include:

i. virus
ii. prokaryotes (bacteria, blue- green algae, cyanobacteria)
iii. eukaryotes (yeasts, molds, animal cells, plant cells, whole plants, whole animals, transgenic organisms).

Examples

i. *agrofood*: food/beverages
ii. organic acids and alcohols
iii. flavors and fragrances
iv. DNA for gene therapy and transient infection
v. antibiotics
vi. proteins (mAbs, tPA, hirudin, Interleukins, Interferons, Enzymes)
vii. hormones (insulin, human growth hormone, erythropoietin, follicle-stimulating hormone, etc.).

Improvements to bioprocess productivity generally come from the following sources: (i) cell lines and (ii) bioprocess control (unit operation, few real-time direct measurements) [36]. The current rather expensive system bioprocesses need to be improved significantly.

Innovation in technological development as well as in production bioprocesses is to be pursued [37]. On the other hand, development in upstream processing (USP) includes various parts: cell line development and engineering, cell clone selection, media and feed development, bioprocess development and scale up [38, 39]. Bioreactor design, process control and monitoring, cell harvesting and the corresponding analytics can be part of the optimization bioprocess as well [37, 38]. These areas are optimized individually and focus on a stable product, high productivity and high quality [37]. Figure 1.12 schematically the different optimization areas and lists the most important variables. These include the establishment of platform technologies, high-through-put methods with approaches based on Quality by Design (QbD) and Design of Experiments (DoE) based on experimental optimizations. Additionally, an integration of modeling and simulation of unit operations as well as the use of mini-plant facilities is applied in bioprocess development [37]. Key steps in upstream process development are

i. *Construct cell line.* Genetically engineer a host organism (mammalian cell line, *E. coli*, etc.) to produce the desired bio-product.
ii. *Clone selection.* Select from many hundreds of clones the most appropriate production cell line based on product quality and productivity.
iii. *Process development.* Optimize production process parameters to maximize productivity, obtain acceptable product quality, and scale-up process to production.
iv. *Process transfer.* Enable information transfer (process and equipment) between process development and production groups to minimize process issues and speed-up commercialization.

Development in downstream processing (DSP) focuses on yield and productivity as well as on purity and bioprocess capacities. An increase in separation (product sources, primary recovery, and purification) efficiency of single unit operations is achieved by expansion of existing facilities and by optimization of existing and alternative processes [37].

1.2.1 Cell Lines

The selection of the expression system is determined by its ability to ensure a high productivity and defined quality criteria. The expression system that is commonly used for the production of monoclonal antibodies or recombinant proteins is Chinese hamster ovary (CHO) cells [38]. The first proteins produced by CHO-derived cell lines were recombinant interferons and tissue-type plasminogen activator (tPA). In 2010, approximately 70% of all recombinant proteins had been produced in CHO cells [39].

High productivity and post-translational processing are the criteria for cell line selection after cell transfection. Other factors, such as growth behavior, stable production, cultivation in serum-free suspension media, amplification, clone selection, and possible

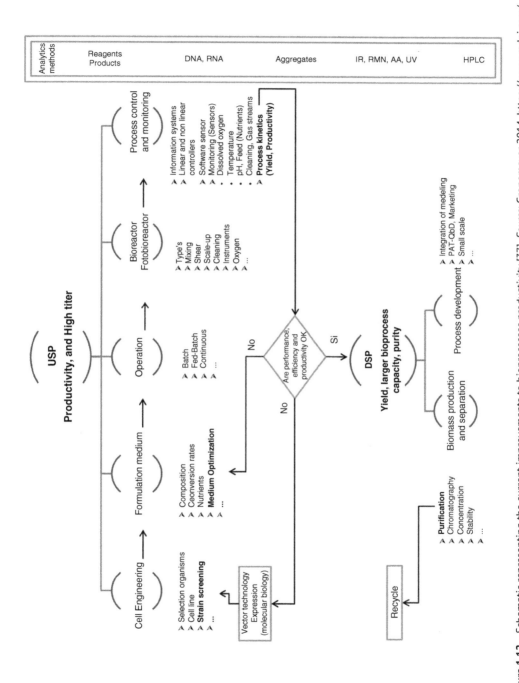

Figure 1.12 Schematic representing the current improvements to bioprocesses productivity [37]. *Source:* Gronemeyer 2014, https://www.mdpi.com/2306-5354/1/4/188. Used under CC-By4.0 https://creativecommons.org/licenses/by/4.0/.

risk assessment are taken into account as well [40]. Provide a robust manufacturing cell culture/fermentation process that has the following performance capabilities

i. steady product quality
ii. stable process yield
iii. regular level of process impurities
   • host cell protein, DNA, and product variants
iv. reliable broth conditions for predictable downstream unit operations performance
   • clarification and primary capture column.

1.2.1.1 Cell Culture Process General

Requirements
i. Phenotypic Stability
   • The cells grow, metabolize, and produce with a predictable profile from batch to batch.
ii. Genetic Stability
   • The cells retain the genetic code for the protein of interest over the population generations from the cell vial to the production vessel harvest.
iii. Axenic Processing
   • The process is free of foreign growth.

The cell clones considered for final production have to fulfill the required product quality, bioprocesses ability and volumetric productivity [41]. Criteria for selection are:

i. cell-specific
ii. growth
iii. volumetric productivity
iv. glycosylation profiles
v. development of charge variants
vi. protein sequence heterogeneity
vii. clone stability [41].

Especially product quality, productivity, and the metabolic profiles of the cells strongly depend on cell culture conditions that include a number of technological trends:

i. improved expression systems
ii. cell lines and genetic engineering
iii. culture media and media optimization
iv. process analytical technologies (PAT)
v. modeling software for process debottlenecking.

In manufacturing facilities, many of bioprocess parameters are being monitored and acquired automatically in real time throughout the entire production train [42].

1.2.2 Microorganism Growth Under Controlled Conditions

The key mission in industrial production bioprocesses is to increase yields and also ensure consistent product. To this end, strain phenotype improvement, bioprocesses optimization,

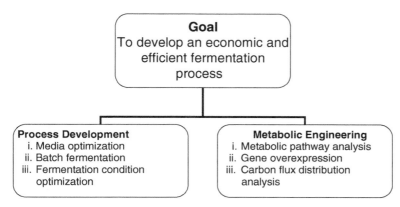

Figure 1.13 Bioprocess operations [43].

and scale-up, the main topics of industrial bioprocesses are aimed at maintaining optimum and homogenous reaction conditions, minimizing microbial stress exposure and enhancing metabolic accuracy.

Microorganism growth is significantly influenced by the sensitivity of microorganisms to their environment. Traditionally variables like **pH**, **dissolved oxygen**, **temperature**, **pressure**, **level**, and **flow** are employed for monitoring, instrumentation, and controlling bioprocesses (see, Figure 1.13). Controlling only the culture parameters like dissolved oxygen (DO), pH, pressure, and temperature are not enough to reduce the variability in the process. In fact, factors like concentration of nutrients, growth balance, intracellular metabolic products, and energy charge are all relevant for understanding the process state [44]. Usually the goal is to make measurements that can be quantified in real time as shown in Figure 1.14.

Microorganisms grow in a wide range of environments:

  i. some like hot environments while others like cold
 ii. grow in an acidic environment
iii. some require high moisture
 iv. some can tolerate high-salt (saline) environments.

Many require the presence of oxygen, but some do not. The overall effect of the underlying mechanisms of cellular regulation dictates a nonlinear behavior in cell culture processes.

Nutrients – change during process:

 i. high energy and N (protein) at first (rapid growth)
ii. very precise conditions later, to maximize yield.

1.2.3   On the Environment for the Microorganism's Growth

The minimum growth temperature is the lowest temperature at which the species will grow, and the maximum is the highest. The optimum growth temperature is that at which it grows best (see Table 1.2). Any nutrient material prepared for the growth of bacteria in a laboratory is called a culture medium. Microbes growing in a container of culture medium are referred

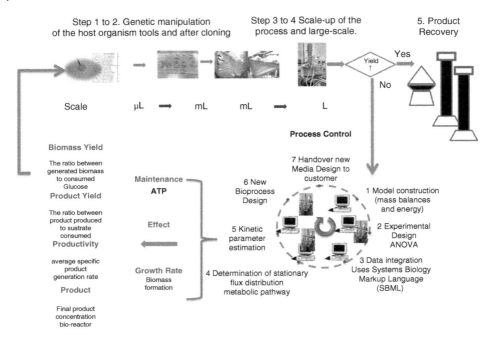

Figure 1.14 Illustrative example for bioprocess monitoring and control.

Table 1.2 The most common requirements for microorganism's growth [45].

Requirements

Temperature	Organisms	Interval	Examples
	Psychrophiles	$-10\,°C$ to $30\,°C$	*Pseudoalteromonas*
	Mesophiles	$25°$ to $40\,°C$	*Escherichia coli*
	Thermophiles	$50°$ to $60\,°C$	*Alicyclobacillus*
	Hyperthermophiles (extreme thermophiles)	$80\,°C$ or higher	*Calyptogena magnifica*
pH	Acidophiles	< 6.5	*Thiobacillus acidophilus*
	Neutrophiles	6.5 to 7.5	*Salmonella*
	Alkaliphiles	> 7.5	*Clostridium*
Nutritional requirements	Carbon: all living things need a carbon source Sulfur: used in protein synthesis		
	Nitrogen: used in protein and nucleic acid synthesis		*Cyanobacteria*
	Phosphorus: used in ATP and nucleic acids		*Bradyrhizobium*
	Trace elements: (Fe, Zn, Cu, etc.) Co-factors and co-enzymes		
Oxygen needs	Obligate anaerobes	$O_2 = 0$	*Actinomyces*
	Facultative Anaerobes	$O_2 \approx 0$	*Desulfovibrio alaskensis*
	Obligate aerobes	$O_2 > 0$	*Staphylococcus species*

to as a culture. When microbes are added to initiate growth, they are an inoculum. To ensure that the culture will contain only the microorganisms originally added to the medium (and their offspring), the medium must initially be sterile (see Table 1.2). Two types of culture medium are available: a broth and a liquid nutrient medium without agar.

Agar: a common solidifying agent for a culture medium. Most bacteria grow best in a narrow pH range near neutrality, between pH 6.5 and 7.5. Very few grow below pH 4.0 (see Table 1.2) bacteria (mostly) at pH = 3 to 8 yeast, pH = 3 to 6, plants pH = 5 to 6 animals pH = 6.5 to 7.5. Microbes that use molecular oxygen are aerobes; if oxygen is an absolute requirement, they are obligate aerobes. Facultative anaerobes use oxygen when it is present but continue growth by fermentation or anaerobic respiration when it is not available.

Facultative anaerobes grow more efficiently aerobically than they do anaerobically. Obligate anaerobes are bacteria totally unable to use oxygen for growth and usually find it toxic. Water is required to dissolve most cell substances that microorganisms use: minerals, ions, gases, and numerous organic compounds. Some bacteria can survive under extremely dry conditions by forming spores, most bacteria grow at a_w 0.85–1.0. So, it is important that the operator understand how the following factors affect the growth of the bacterium:

  i. oxygen utilization
 ii. sludge age
iii. dissolved oxygen
 iv. mixing
  v. pH
 vi. temperature
vii. nutrients.

1.2.4 Improving the Productivity for Specific Metabolic Products

The range of bio-products that can be produced has expanded, in part due to notable advances in fields adjacent to metabolic engineering and extensive databases of gene expression, therefore metabolic engineers must consider many factors in the development of microbial catalysts [46]:

 (i) metabolic route
 (ii) genes encoding the enzymes in the metabolic pathway
(iii) gene-control systems
(iv) computational methods for debugging and debottlenecking the novel pathway.

The most commonly used modeling framework for metabolic networks is the flux balance analysis (FBA). Several FBA-based algorithms have been designed to predict the genetic manipulations required to maximize the production of the desired chemical (product yield and productivity) [47, 48] (Figure 1.15).

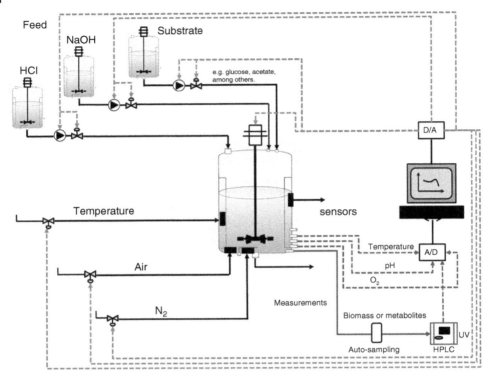

Figure 1.15 Efficient fermentation process.

1.3 Bioprocess Control

1.3.1 What Is a Bioprocess?

Bioprocess engineering deals with the design and development of equipment and processes for the manufacture of products such as food as feed, pharmaceuticals, chemicals, polymers, and paper from biological materials or microorganisms. Bioprocess is a combination of mathematics, biology, and industrial design which consists of various spectrums like designing of bioreactors, study of fermenters (mode of operations), etc. (see, Figure 1.14). Moreover, bioprocess engineering is also devoted to environmental waste treatments of sustainable energy via biofuels production.

Bioprocess engineering also deals with the study of various biotechnological processes used in industries for large-scale production of biological product for optimization of yield in the end-product and its quality. Bioprocess operations use microbial, animal, and plant cells and components of cells such as enzymes to manufacture new products and destroy harmful waste [43, 47].

Humanity's ability to harness the capabilities of cells and enzymes has been closely related to advancements in microbiology, biochemistry, and cell physiology stages. Visions of sophisticated medicines or protein treatment (as insulin), cultured human tissues and organs, biochips for new-age computers, environmentally-compatible pesticides,

Table 1.3 Most bioprocesses consist of some general phases that are conducted as a sequence of batch steps [40, 49].

Step	Name	Description
1	Genetic manipulation of the host organism tools	i. stability of the constructed strains ii. level of expression of the desired product
2	After cloning (small-scale) 150 ml to 1 l	Parameters: i. medium composition ii. pH iii. temperature iv. specific productivity v. product yield
3	Scale-up (1 or 2 l)	Instruments and economic feasibility for measuring and adjusting: i. temperature ii. pH iii. dissolved-oxygen iv. stirrer speed
4	Pilot-scale (100–1000 l)	Parameters: i. geometry of the reactor ii. method of aeration and mixing iii. impeller design iv. process-control
5	Product recovery (downstream, 80–90% of the total processing cost)	Operations: i. filtration ii. centrifugation iii. flotation iv. mechanical disruption v. solvent extraction vi. chromatography vii. membrane filtration viii. adsorption ix. crystallization and drying

Source: Reproduced with permission of Elsevier.

and powerful pollution-degrading microbes herald a revolution in the role of biology in industry. Biological systems can be complex and difficult to control; nevertheless, they obey the laws of chemistry and physics and are therefore amenable to engineering analysis [49] (Table 1.3).

1.3.2 Bioprocess Monitoring and Control

Among the bioprocesses and bioengineering, the fermentation processes are considered complex for controlling and monitoring due to the intricate nature of biological systems (e.g. medium is a mixture of simple and complex components) and their interaction with the surrounding physical and biochemical environment [44].

Moreover, fermentation varies from product to product being formed in different phases of the bioprocess. In addition to control the culture parameters (like DO, pH, pressure, or temperature), some factors such as concentration of nutrients, growth balance, intracellular metabolic products, and energy load are all relevant for understanding the process state.

In fact, an essential requirement is that the sensing techniques employed need to be rapid not only to facilitate a deeper understanding of the process but also to utilize them for process control including physiological stages during the bioprocess [50]. In production, only critical parameters are usually available from measurements to enable proper control of the process and ensure high quality and yield; however, in practice, less suitable sensors and tools for online monitoring have not allowed this idea to be widely implemented [51].

Figure 1.15 shows some of the more common bioreactor operational modes used in bioprocessing, which are prerequisites to appreciate the differences in modeling approaches. Batch culture is the only closed culture mode. Fed-batch, continuous, and perfusion culture are semi-continuous or continuous modes of operation as compared with repeated batch and medium exchange, which are discrete culture modes of operation. Additional techniques are under development and can include combinations from dielectric and Raman spectroscopies, cell counter systems and fluorescence flow cytometry [50]

For industrial developments the central and manifold objective of the control is the realization of economic interest in assuring the named operational stability, process reproducibility, and increased product yield together with maintaining rigorous safety and the implementation of good manufacture practice (GMP) or environmental regulations, important requests in modern bio-manufacture imposed by the product quality improving needs [52].

There are two main obstacles for controlling bioprocesses

i. Modeling complications
   - how to account for the numerous factors that influence the biochemical reactions (including the growth)
   - nonlinear and non-stationary models
   - specific configuration and changes at the physiological stages.
ii. Measuring limitations
   - absence of cheap and reliable sensors for the key process variables
   - crossing methods for the inference though indirect measurements.

The key monitoring parameter that many bioprocess engineers rely on is viable cell density. This is a compound parameter arising from measurement of the cellular concentration and the total cellular viability. Viable cell density is the mathematical product of these two primary pieces of information [52].

Bioprocess advancement is determined by the living cell's capabilities and characteristics, the bioreactor performance as well as by the cultivation media composition and the main parameter's evolution. The high metabolic network complexity inside the cells often determine very sophisticated, nonlinear growth and product formation kinetics, with further consequences on the bioprocess behavior, but at the same time on the product quality and yield. What we can measure routinely today are the operating and secondary variables such as the concentrations of metabolites that fully depend on primary and operating variables.

In comparison to other disciplines such as physics or engineering, sensors useful for in situ monitoring of biotechnological processes are comparatively few; they measure physical and chemical variables rather than biological ones. The reasons are manifold but biologically relevant variables are much more difficult and complex than others [53]. Another important reason derives from restricting requirements,

i. sterilization procedures
ii. stability and reliability over extended periods
iii. application over an extended dynamic range
iv. non-interference with the sterile barrier
v. insensitivity to protein adsorption and surface growth
vi. resistance to degradation or enzymatic break down.

The main bioprocess control attributes are: handling of off-line analyses, recipe and scheduling, high level overall control, state and parameter estimation, simulation, prediction, and optimization.

1.3.3 Stability of Bioprocess

In biotechnological applications, the appearance of multiple steady states could lead to undesirable situations for bioprocess operation, consisting usually of many species and reactions that can also give rise to complex reaction behaviors. It is important for the bio-chemical engineer to be able to identify bio-chemical systems that have the capacity to exhibit multiple steady states, since such identification helps to design more efficient and safer bioreactors [54]. Even for the simplest bioprocess, the chemostat, the formal mathematical analysis might involve unification of criteria or the analysis of a study case. The unification criteria require that, for instance, conditions for generic growth rates include phenomena such as substrate, biomass, and product inhibitions [55].

The stability of a system implies a tendency to reach a desirable condition when it has been disturbed (perturbed). In general, stability is the property of a system that influences it to return to its original position. In other words, for example, if a system is in a static equilibrium condition and it is subjected to disturbance from such a condition, if it then comes back to its initial position then it is found to be stable, otherwise it is termed unstable. Any kind of nonlinearity in the system as well as time delay become sources of instability and bring down the system performance. Any unstable system is undesirable. Hence, stability analysis is a very important concern to be addressed. Thus, stability analysis becomes most relevant from the operative viewpoint, which involves the control system design and becomes fundamental for engineers and scientists.

The concept of stability is central to the analysis of dynamical systems. Depending on the particular type of system – for example, whether it is linear or nonlinear, whether it has inputs or not – but in all cases, however, the underlying concept is that a system is said to be stable if small perturbations whether in the initial condition or due to external stimuli, do not give rise to large sustained changes in the behavior of the system.

In order to incorporate an illustrative notion, let us suppose that the system has equilibrium $\bar{s}$. We said that the equilibrium is locally stable if each initial condition s_0 that is near $\bar{s}$ gives rise to trajectories that stay close to the equilibrium (initial conditions) belonging to a neighborhood of $\bar{s}$. Note that stability does not necessarily require that the trajectory tends to equilibrium as $t \rightarrow \infty$, i.e. asymptotical stability. Complementarily, the trajectory can be convergent at finite time. For example, if the convergence rate is governed by an exponential time function, and holds onto the equilibrium $\bar{s}$ a long time, then it is called a local exponential stability. In addition, as the initial conditions belong to a set larger than the

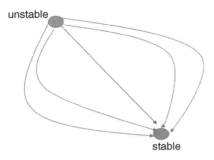

unstable

stable

Figure 1.16 Concept of stability.

equilibrium neighborhood then it is a candidate to be called global stability. Where equilibrium is unstable, nearby trajectories will diverge from the steady state (see, Figure 1.16).

$$\frac{d}{dt}x(t) = f((x)) \tag{1.20}$$

It should be emphasized that stability is not a property of a system but is rather solutions of the dynamical system. That is, each isolated steady state z^* is a solution in the sense that $\dot{z} = f(z^*) = 0$ Moreover, any closed orbit $z(t)$ satisfying the differential equation is a solution as well. These kinds of solutions can be stable or not, depending on the vector field $f(z)$. The nonlinear systems can have multiple equilibria, with different stability properties (for further details see the Chapter 2).

The notions of stability and instability of solutions as equilibria can be illustrated by considering the behavior of balls on an undulating slope. For balls rolling on the slope depicted here, there are three possible equilibria, corresponding to the valley bottoms (stable) and the top of the hill (unstable). A ball balanced perfectly on the top of the hill (b) will remain there, but the slightest perturbation will cause it to roll away. Conversely, a ball in a valley bottom ((a) or (c)), if displaced by a small amount, will return to its resting place (see, Figure 1.17). Many bioprocesses, including high purity distillation columns, highly exothermic biochemical reactors, and batch bioreactors frequently present multiple steady states or multiplicity behaviors due to the complexity of the relationships between the operating conditions of bioreactors and the product properties.

Toward a formal stability study, a suitable mathematical model of a given system is needed. Any nonlinear system can be linearized around an operating point and the stability study can be performed. For analysis and design purpose, the stability can be classified into

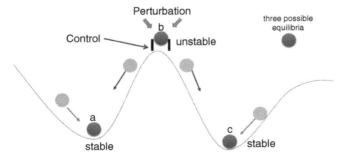

Perturbation

b

Control — unstable

three possible equilibria

a

stable

c

stable

Figure 1.17 Scheme of stability and unstable of equilibrium.

absolute stability and relative stability. Absolute stability dictates the condition of whether the system is stable or not, once the system is found to be stable, it is of interest to infer how stable it is, and the degree of stability is a measure of relative stability.

1.3.4 Basic Concepts and Controllers

The objective of process control is to achieve production of the desired product efficiently and effectively in a safe and profitable manner. The quality of the final product from a bioreactor depends mainly on the control loop employed to monitor and control the microbial growth based on the reference input. Controllers operate on the error signal of the measured variable and generate a control signal that acts on a final control element or the actuator. This implicates a material change in the process to ensure the controlled variable is at the desired set point, irrespective of process disturbances. In the bioindustry, bioprocesses are subject to a number of local and/or supervisory control structures. Local controllers are used to get the set-point control of different physical/chemical parameters (e.g. temperature, pH, and dissolved oxygen concentration), while supervisory control is necessary for optimizing the feed in a fed batch process or the dilution rate in a continuous one [56]. For example, the controlled variable, e.g. substrate concentration or dissolved oxygen, is measured and influenced by a manipulated variable, e.g. feed rate, agitation, nitrogen source, to ensure it is always at the desired value.

i. *Simple indirect feedback methods.* Nutrients (indirect variable) are fed to the bioreactor by an on–off controller when a direct (on-line measured) variable deviates from its set point (temperature, level, pH, and dissolved oxygen).
ii. Feedforward strategy based on prior process knowledge (semi-batch operation).
iii. *Direct feedback.* A dilution rate can be directly controlled by nutrient feeding when it is measured on-line by sensors inside or outside the bioreactor.
iv. *Feed control by state estimation.* The estimation of key-process parameters from on-line measurements can be applied and the control is based on the dynamic of the biomass or the substrate concentration [56].

Concepts

i. *Controller.* To maintain the desired conditions in a physical system by adjusting selected variables in the system. The controlled quantities (or controlled variables) are those streams or conditions that the practitioner wishes to control or to maintain at some desired level. The input–output relationship represents the cause-and-effect relationship of the process.
ii. *System.* A system is a combination of components that act together, to meet a goal, there are open (continuous) systems in which matter is transferred by the system boundary, or closed (by batch) where there is no transfer during the time interval of interest, a control approach defines it as any process that produces a signal transformation.
   The concept of the system allows us to raise the understanding of reality in two major stages:
   - identifying physical systems
   - establishing the rules or laws that describe them.

iii. *Process*. Is any operation to be controlled (example: physical, chemical and biological process, in a wastewater treatment plant).

iv. *Set point*. Reference values also known as the desired value or reference input.

v. *Error*. The difference between the set-point and the measurement, controller responds to error:

$$e = x - x_{set-point} \tag{1.21}$$

Error e depends on state x (via sensors y).

vi. *Plant*. Any physical object to be controlled. The purpose of the plant is executing a particular operation. For example, a plant's purpose may be moving an object to a particular position.

vii. *Open-loop*, is a control system which utilizes an actuating device to control the process directly without using feedback.

viii. *Closed-loop* is a control a system which uses a measurement of the output and feedback of the signal to compare it with the desired output (reference or command). Advantages of using feedback:

- it compensates for disturbances
- better stability performance
- better bandwidth performance
- better sensitivity performance
- better performance of global profit.

ix. *Measured variable*. Measured value of process variable being controlled.

x. *Final element*. The actuator used to control flow of a fluid.

xi. *Controller signal*. Signal going out from the controller to the actuator.

xii. *Sensor*. Device used to send the signal to the control room.

xiii. *External disturbances*. Variables that could affect the controlled output.

xiv. *Manipulated variable*. There is an associated manipulated quantity; in such, cases the flow rate of the stream is often manipulated through the use of a control valve.

xv. *Controlled variable*. The controlled quantities (or controlled variables) are those streams or conditions that the practitioner wishes to control or to maintain at some desired level. These may be flow rates, levels, pressures, temperatures, compositions, or other such process variables for which the practitioner also establishes some desired value, also known as the set point or reference input.

xvi. *Inputs*. The input u is, for example, the flow rate or heater power. Not only the reference signal (desired value of the system output) conforms to an input variable, there are also some undesired signals, such as some external disturbances. In turn, these channels exist as two different types: those accessible to manipulation and those inaccessible to manipulation. The input channels accessible to manipulation can provided information to the system.

xvii. *Outputs*. The output y is the temperature. These channels exist as two types, those that provide measurable information and those that are inaccessible to measurement (which can be observed in certain circumstances, this is measured indirectly).

xviii. *State variable* is defined as a set of variables that allow prediction of the evolution of the system in complete absence of disturbances.

xix. *Four basic operations in any control system*.

Measurement	• The measurement of the variable to be controlled is made by combining a sensor and a transmitter.
Decision	• Based on the measurement the control algorithm makes the decision to keep the variable to be controlled in the desired condition, this involves the sensor, transmitter and an interface.
Action	• According to the decision the controller develops an action in the system made generally by a final control clement such as a pump or valve.
Monitoring	• All these operations will allow the user to perform an online monitoring for the purpose of diagnosing faults during the process, modifying the control objectives to correct the effect of any disturbance to the system (Interface-User).

xx. *Dynamic system.* This is the one that generates data that change with the passage of time, that is, they possess certain dynamics. Dynamic systems are systems whose internal variables (state variables) follow a series of temporal rules. They are called systems because they are described by a set of equations that are time dependent, either implicitly or explicitly. The study of dynamic systems can be divided into three sub-disciplines:

- *Applied dynamics.* Process modeling through state equations that relate past states to future states.
- *Mathematics of dynamics.* Focuses on the qualitative analysis of the dynamic model.
- *Experimental dynamics.* Laboratory experiments, computer simulations of dynamic models.

Dynamic system classes

- Isolated. *They do not interact with their environment.*
- Natural. *Unaffected by human intervention.*
- Artificial. *Created by man.*
- Physics. *They involve matter and energy.*
- *Not* physical. Thoughts.

xxi. *Linear systems.* A system is called linear if the superposition principle is applied. This principle states that the response produced by the simultaneous application of two functions of different inputs is the sum of the two individual responses. Thus, for the linear system, the response to several inputs is calculated by treating one input at a time and adding the results

xxii. *Nonlinear systems.* A system is nonlinear if the superposition principle is not applied. Therefore, for a nonlinear system the response to two inputs cannot be calculated by treating each one separately and summing the results, represented by the system of nonlinear differential equations.

Controllers employed a range from simple on–off type to proportional (P), integral (I), derivative (D), and proportional integral derivative (PID) controls and expert systems.

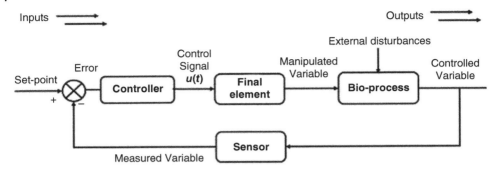

Figure 1.18 Overall scheme of bioprocesses and basic control structures.

A scheme of classical feedback control is shown in Figure 1.18 and a mathematical expression for the classical PID becomes:

$$u(t) = \overbrace{K_p e(t)}^{Pr\,oportional} + \overbrace{K_i \int e(t)dt}^{Integral} + \overbrace{K_d \frac{d}{dt} e(t)}^{Derivative} \tag{1.22}$$

K_P, K_d, and K_i are the tuning parameters of the controller that can be adjusted by varying the dynamics of the control loop. Feed rates can also be adjusted based on an optimal objective function derived on-line or off-line. The objective function target is to increase productivity or maximize operating profits. Sometimes a combination of optimal and feedback controls is used or advanced design approaches can be performed to deal with specific issues.

Sometimes it is desirable to compute a performance index for the classical feedback control. The underlying ideas about performance indexes looks to measure, at least, two issues: the difference between current and desired behavior and the magnitude of the control effort to reach the desired behavior. That is, the index is computed along with the time of the control execution. Typically, an index can take the following form:

$$I = \overbrace{\frac{1}{T} \int abs(y - y^*)dt}^{Convergence\ error} + \overbrace{\frac{1}{T} \int abs(u - u^*)dt}^{Control} \tag{1.23}$$

where *abs* denotes the absolute function, y^* the desired behavior of the system output u^* stands for the nominal control whereas y and u are the system output and control input, respectively. T corresponds to the overall execution control time. Depending on the requirements, the *abs* function can be substituted by a quadratic term or any norm.

1.3.5 Advanced Control Schemes: Multivariable Control, Robust, Fuzzy Logic, Model Predictive Control, or Others

Since bioprocesses involve living organisms, they often experience nonlinear behaviors that may include output multiplicity, bifurcations, chaos, unstable dynamic response to disturbances, and changes in system parameters; all these phenomena can lead to instability and, ultimately, affect the yield of production. For this reason, the application of traditional linear controllers is limited since they are unable to cope with the highly nonlinear behavior

Table 1.4 Summary: model controllers applied in bioprocesses [59].

Type	Process	Note	Ref.
Nonlinear predictive control	Fermentation	Simple local state-space model structures are developed for each regime	[60]
Neural network	Semi-batch	The process is performed in a semi-batch pilot-plant reactor	[61]
Adaptive control	Fed-batch	The robust controller uses minimal process knowledge and minimal measurement information	[62]
Adaptive control	Batch-continuous	Design of an adaptive nonlinear proportional–integral (PI) controller to regulate the dynamics of an aerobic wastewater bioreactor, where a mathematical sketch of proof of the convergence of the control scheme is provided	[63]
Fuzzy control	Automating the bioprocesses	Applications of fuzzy control to industrial biological processes	[64]
Model-based geometric control algorithm (MGA)	Fermentation	Comparing the performance of IMC, PI, and MGA	[65]
Robust H-infinity for blood glucose regulation	Type 1 diabetes mellitus	Synthesis and analysis of feedback controllers with robustness properties in face to parametric uncertainties and physiological biosignaling, avoiding nocturnal hypoglycemic events, and accounting exercise	[66–70]
Positive theory on systems with positive control	Type 1 diabetes mellitus	These approaches exploit the fact that state variable in bioprocess evolve only at positive values and the control input is positive as well	[71, 72]
Sliding-mode for extremum seeking	Continuous configuration for bioreactors	Strategies to find automatically the maximum yielding toward biogas production or maximum conversion	[73–78]
Robust adaptive approaches	Wastewater treatment to DQO regulation	Adaptive approach without projection algorithms and experimental proof-of-concept for feedback–feedforward control and robustness in face to uncertainties in the model	[79]

of biological processes. Aside from this, incidental external and internal disturbances in a bioreactor may result in bioreactor failure.

Therefore, there is a strong financial motivation to develop the finest control scheme that would facilitate rapid start-up and stabilization of continuous bioreactors subject to redundant disturbances. In the literature, a variety of methods have been discussed to implement a robust controller for the bioreactor operating at single or multiple steady states. To deal with uncertainties and nonlinearities, significant results can be obtained with sliding mode controllers (SMC), which represent a simple approach to robust control [57].

Sivakumaran et al. have discussed the recurrent neural network (RNN) based modeling method for a nonlinear bioreactor operating at a single steady state, and implemented a nonlinear model predictive controller [58]. Giriraj Kumar et al. have discussed a genetic-algorithm (GA) based PI controller tuning for a bioreactor operating at stable steady state. A bioreactor operating at an unstable steady state was widely discussed by the researchers due to its complexity and instability [59].

In the control literature, regardless of the considerable progress in advanced process control proposals such as SMC, model predictive control (MPC), and internal model control (IMC), PID controllers are still widely employed in industrial control systems because of their structural simplicity, reputation, robust behavior, and easy implementation (see, Table 13). Along with the system's stability, it also satisfies the primary performance such as smooth reference tracking, efficient disturbance rejection, and measurement of noise attenuation criteria [57] (Table 1.4).

1.4 Process Measurements

1.4.1 The Drawback for Monitoring Bioprocess

Generally speaking, biological systems are influenced by different process variables that have a direct influence on cell metabolism. Sensors for these variables are (typically) inserted into specially designed ports on the bioreactor. As bioreactors increase in size (i.e. in the industry field), the mixing problems become common and probe location becomes problematic.

Process monitoring helps to track the process and reduce this variability. Further, to achieve process optimization and control, it is desirable to maximize product production for a given substrate and other consumed resources. For any bioprocess, the real-time view of key process variables that affect the critical quality attributes of the product being produced is crucial. Monitoring helps increase productivity and yield, saving costs while improving efficiency. Further, process monitoring provides tools for risk management, diagnostics, and continuous improvement of processes and addresses the key requirement of PAT framework. Classical methods of analysis of samples from bioprocesses include chromatography and mass spectrometry. However, efforts are progressing to integrate and have rapid online analysis of biochemical parameters and, in this context, biosensors are important [44, 64].

Expected features and characteristics of monitoring techniques for bioprocesses

   i. reliable
  ii. sensitive
 iii. non-invasive
 iv. non-destructive
  v. stable, rapid, and simple to use
 vi. not compromising sterility
vii. biologically inactive.

1.4.2 Primary on-Line Sensor (e.g. Dissolved Oxygen, Temperature, Culture pH, Pressure, Agitation Rate, Flow Rates, Redox, CO_2, and Others)

The standard direct physical determinations in bioprocesses are:

i. *Temperature.* Sterilization (15 minutes at 121–124 °C- 200 kPa), concentration (effect of temperature on the kinetics of cell growth), and purification.
ii. Pressure (gas partial pressures [CO_2, O_2, CH_4, etc.], dissolved gases [O_2, CO_2, CH_4, etc.], concentrations of volatiles [methanol, ethanol, lactate, acetone]).
iii. Power input is associated with hydrodynamic stress (impeller distance), which may affect cell growth and/or productivity of shear-sensitive production organisms(control agitation, gas flow, and vessel temperature).
iv. Foam control in fermentation bioprocess.
v. Gas and liquid flow (measure flow rates).
vi. Weight and low molecular weight contents in perfusion processes.

The regular chemical determinations are:

i. pH (controlled by the addition of appropriate quantities of alkaline or acid solutions).
ii. Dissolved oxygen (the dissolved oxygen concentration does not fall below a specified minimal level [anaerobic-aerobic systems]).
iii. Redox potential (microbial cultures).
iv. Ion-specific sensors (ion-specific sensors; Mg_2^+, NH_3^+, Ca_2^+, K^+, etc.).
v. Enzyme electrodes (the specific enzyme is immobilized on a membrane in close contact to a pH electrode; a computer-controlled process-FIA system for monitoring).
vi. Microbial electrodes (immobilized whole cells have been used also for the determination of sugars, acetic acid, ethyl alcohol, vitamin B, nicotinic acid, glutamic acid, and cephalosporin; fluorescence sensors for monitoring and controlling).

In most bioprocesses there is a need for pH monitoring and control if maximum productivity is to be obtained. The pH can change, depending on the characteristic pH trend evolution (set-point). In most bioprocesses, the polarographic electrodes are more commonly used in pilot or production bioreactors. The mass spectrometer can be used for in-line analysis since it is very versatile, but expensive. The fluorimetric methods are very specific and rapid, but their use in bioprocesses is quasi-limited nowadays, but expensive. Hence, the measurement of NAD would be an ideal method for continuous measurement of microbial biomass concentration [44, 65].

1.4.3 Primary in-Line Sensor

Importance of biosensors in bioprocesses can be gauged by the fact that many compounds of interest such as amino acids, antibiotics, biomass, proteins, sugars, alcohols, etc. can be quantified quickly by biosensors. Amino acids are important both as a by-product of fermentation and as a factor relevant for production. Monitoring of biosynthesis of penicillin, sucrose, glucose, phenol, tyrosine, antibiotic, and based on enzyme thermistors has been reported (lactamase as the enzyme). For instance, determination of L-phenylalanine by an enzyme-based biosensor using L-phenylalanine ammonia lyase followed by potentiometric detection (DNA hybridization). Another important amino acid is glutamine which is of significance in mammalian cell culture, and can be quantified (amperometric biosensor) by using glutaminase and glutamate oxidase employing chemiluminescence detection (emission of light as a result of a chemical reaction). DNA biochips exploit hybridization of target DNA strands with fluorophore-labeled cDNA probes that are optical detection (electrochemical detection of hybridization). With high integration capacity of biochips (DNA microarray, protein microarray, and microfluidic chip), an array of immobilized DNAs can accommodate all the required gene expressions that need to be monitored during recombinant-protein production [44, 65].

1.4.4 Process Analytical Technologies (Gas Analysis, Spectrometers, Infrared, HPLC, PCR, and Others)

Spectroscopic sensors enable the simultaneous real-time bioprocess monitoring of various critical process parameters including biological, chemical, and physical variables during the entire biotechnological production process [80]. The quantification and control of the biomass has been constantly advanced on account of the importance of this particular state variable in a variety of bioprocesses, since the introduction of the PAT initiative by the Food and Drug Administration (FDA) in 2004. So far, PAT applications have been related more to small molecule synthesis, formulation, and fill–finish processes, however, the four main principles of the initiative can also be applied to monitoring and control production processes of biologicals such as recombinant proteins. As pointed out by Gnoth et al.: PAT is used to analyze the impact of measurable and controllable state variables on product properties that are not directly measurable in real time. Because of the complexity of the biological system, the access to critical bioprocess variables is limited or the sensors are not robust enough for the required conditions (e.g. pH, T, saline, acid, etc.) in bioreactors [81]. The application of on-line measurements in the bioprocess of high-performance liquid chromatography (HPLC), allows for the measurement of substrate and product concentrations during fermentation. A disadvantage of on-line HPLC is the time delay between sampling and determination of the concentration of the observed process variable. Therefore, it is necessary to consider a law of robust predictive control of these external disturbances considering the time of delay [82, 83]. In addition, bioprocess states can be estimated by using sensitive on-line software sensors in combination with phenomenological models for the estimation of state variables.

Consequently, the combination of soft-sensors with multivariate data analysis (signal) enables bioprocess monitoring and control. In fact, NIR-spectroscopy has been used for bioprocess monitoring and fault diagnosis, as well as Raman spectroscopy. However, both methods are not as sensitive as fluorescence-spectroscopy, a non-invasive technique that enables on-line measurements in media and metabolite concentrations (see, Figure 1.18 and 1.19) [82, 83, 90] (Table 1.5).

The biosensor is based on a biological receptor, and can be used to measure the concentration of different substrates and metabolites in the culture medium [65].

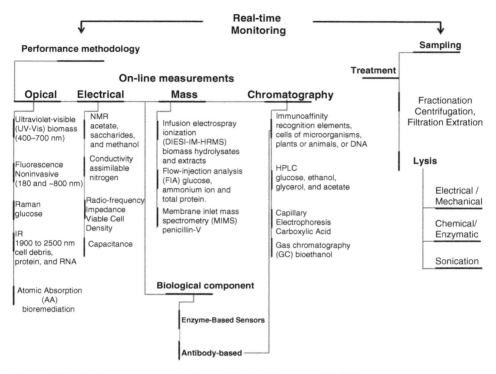

Figure 1.19 On-line measurements for monitoring bioprocess [83–89].

Table 1.5 Feedback control strategies with focus on the monitoring technique employed.

Control law	Monitoring (PAT)	References
Model-reference adaptive control	PAT framework, dissolved oxygen is the only required online measurement	[91]
Feedforward control strategy in aerobic fed-batch cultures	In-line lactose concentration by colorimetry and on-line monitoring of DO levels for estimating the specific growth rate	[92]
Growth was maintained at a point just above critical by regulating ethanol concentration in the bioreactor	On-line ethanol measurement in gas phase and HPLC validation points, off-gas composition was monitored with an infrared analyzer for CO_2	[93]
The specific growth rate of a recombinant *E. coli* strain was controlled during fed-batch cultivations by estimating the specific growth rate from the measured heat flow produced by the cells	Calorimetric methods to high cell density cultivation (HCDC)	[94]

1.4.5 Software Sensor (e.g. Cell Mass Estimation Via Complex Medium, Primary Carbon Substrate, Concentration Product of Line, Metabolites, Sensor to Computer Via Wireless)

In particular, for biotechnological processes the optimal performances depend on available information. Because these variables are frequently associated with the process output quality, they are very important for bioprocess control and monitoring. For this reason, it is of great importance to deliver additional information about bioprocess variables, which is precisely the role of the software sensor [95].

However, the development, and especially the implementation, of advanced monitoring and control strategies on real bioprocesses is difficult because of the absence of reliable instrumentation for the biological state variables, i.e. the substrates, biomass, and product concentrations; for example, the required quality of monitored data, precision data, time delay, frequency of sampling, are a function of the accuracy of bio-sensors, usually requiring more sophisticated measurement devices, which can have several drawbacks, e.g. sterilization, discrete-time (and often rare) samples, relatively long processing (analysis) time, in many cases the state variables, are not on-line (and real-time), this is related to high cost sensors and extreme operating conditions, these facts together with the nonlinearity and parameter uncertainty of the bioprocesses requires an enhanced modeling effort and modern state estimation and identification strategies [96, 97].

A solution to these latter problems can be found through the design of software sensors, this estimation approach is done by combining some available measurement devices to provide signals, such as dissolved oxygen, glucose, pH and temperature and a mathematical model, to provide continuous time estimates of non-measured variables on-line, the estimation algorithm is called a state observer [97]. A software sensor is defined as a grouping of the words "software" because the models are generally computer programs (algorithm), and "sensors" because the models are delivering similar information as their hardware equivalents [97].

Virtual sensors supply indirect measurements of critical conditions (that, by themselves, are not in real time measurable) by combining sensed data from a group of direct measurements sensors and mathematical modeling. The estimation techniques for bioprocesses must be developed in agreement with the specific conditions of the process, according to the specific case. In variable state reconstructing, it is unavoidable that some deviation occurs between the trajectories of a real system and the predictions of a mathematical model. Moreover, there are disturbance signals in virtually all engineering applications and the combined effect of these problems generates the need to study the design of observation techniques that are robust against disturbances and model uncertainties [95, 98] (Figure 1.20–1.24) (Table 1.6).

Table 1.6 Advantages and disadvantages of different methods for software sensor.

Approaches	Input process	Advantages/disadvantages	Example [95]
Mathematical modeling (empirical equations)	Input–output	Approximate models/cannot be viable in the presence of process uncertainties of modeling and measurement noise	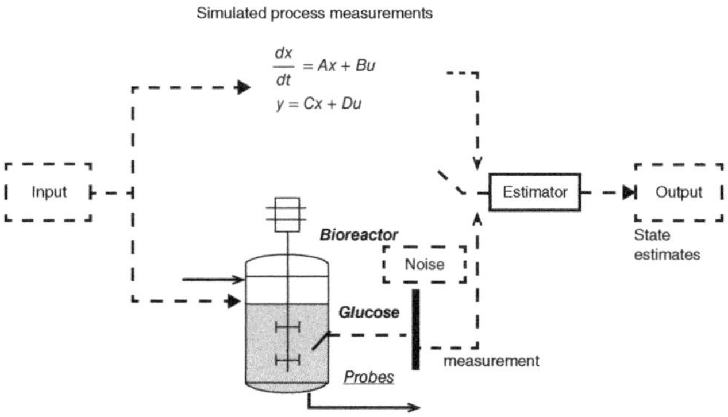
Observer based	Mathematical model	Provide state estimation in deterministic systems/stochastic disturbances	

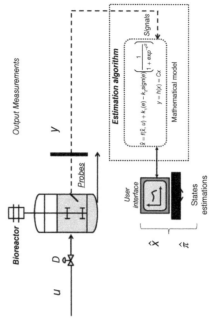

Figure 1.20 Balance equation based methods.

Figure 1.21 Observer based on mathematical model.

(continued)

Table 1.6 (Continued)

Approaches	Input process	Advantages/disadvantages	Example [95]
Kalman filter	Mathematical model	Estimation of states and parameters/convergence require much time and effort the computer	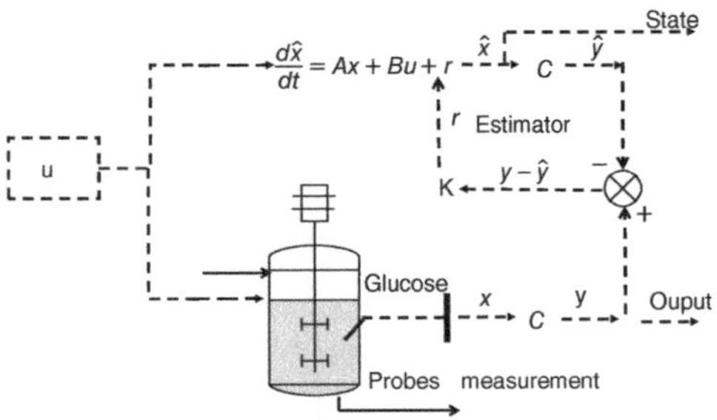
Neural networks	Heuristic data in the form of empirical nonlinear correlations	Provide better performance with noisy and incomplete data/poor generalizations capability outside the training range	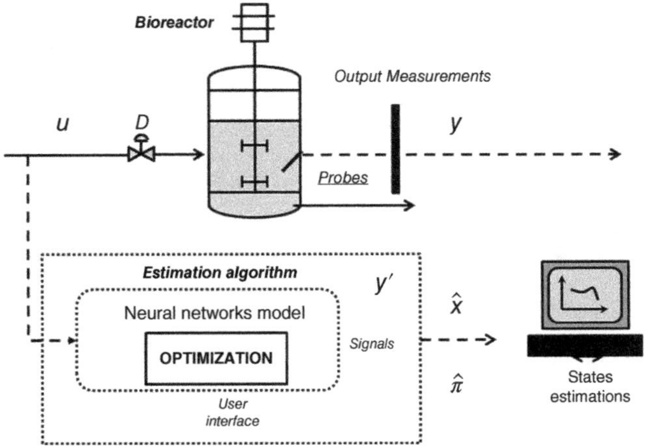

Figure 1.22 Kalman filter.

Figure 1.23 Neural networks.

Fuzzy

Heuristic knowledge in the form of production rules

Useful alternatives for processes that are complex, imprecise and vague/require good understanding of a process to set up a complete and consistent rule base.

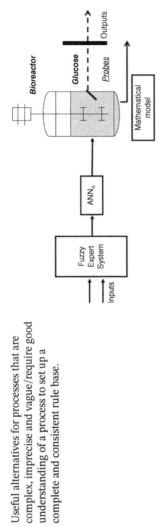

Figure 1.24 Fuzzy reasoning.

1.5 Dynamic Bioprocess Models

1.5.1 Bioprocess Modeling for Control Purposes

Bioreactors play a vital role in green industries to produce important chemical and biochemical compounds. In this system, living organisms also known as microbes (plant, animal, and bacterial cells) are converted into marketable products.

Actually, there is no theoretical basis for any of the reaction kinetics proposed currently in the literature, due to the complexity of the reaction mechanisms that occur in biological systems and to an exact stoichiometric relationship that allows theoretical limiting of the reaction rates similar to a chemical process. Therefore, models must be developed on a type of environmental conditions basis to depict the approximate phenomena under the expected bioreactor conditions. Naturally, a complex mathematical model attempting to encompass as many conditions should result in the greatest predictive capability (increased time and computational effort), but in practice this is usually not practical. As a phenomenological system, bioprocess models rely entirely on experimental data to both fit and validate a proposed kinetic mechanism.

Depending on the measurement technique, measurements can have a high degree of variability of the model proposed:

- Chemical Oxygen Demand
- Various Soluble Organics
- Nitrogen
- Ionic Species
- pH
- Gas Composition.

The mathematical modeling of bioreactors presents an impressively large challenge due to the many high-degree nonlinearities (especially in the complexities of the kinetic proposed), as well as the operation regimen (batch, continuous, and fed-batch), the changes in their culture environment (stress, fluxes of matter and energy, pH, T, pressure, metabolites), and driven by their genetic information, living cells alter the kinetic rate of their due to changes in their metabolic pathways. Growth of bacteria, either single species, mixed cultures, or complete ecosystems can be described in terms of dynamical systems [99].

The kinetic model is represented by the biochemical reactions, including enzyme activity. These are well developed and analytically examined reactions of enzyme catalysis (Michaelis–Menten, Higgins, and Reich) and other local models based on ODEs. Analytical and numerical examination of these models allowed the conditions for the emergence of qualitatively new regimes to be formulated: multi-steady-state, self-oscillating, and quasistochastic in the chains of metabolic reactions [100].

It is generally assumed that mathematical modeling helps us to understand the internal and external dynamics observed in the processes to formulate predictions about their different conditions. The term model means different things in different areas, here we will discuss that which prevails over disciplines ranging from experimental bioprocess, through chemistry up to experimental physics.

The general dynamical model in a stirred tank bioreactor can be described by the following mass balance equation written in a matrix form as:

$$\frac{d\xi}{dt} = K\mu(\xi, t) - D\xi + F - Q \tag{1.24}$$

in which, ξ is a vector representing the n state component concentrations ($\xi \in \mathfrak{R}^n$), μ is the growth rate vector corresponding to m reactions ($\mu \in \mathfrak{R}^m$), K is the matrix of yield coefficients ($K \in \mathfrak{R}^{n \times m}$), F is the vector of feed rates and gaseous outflow rates ($F \in \mathfrak{R}^n$), D is the dilution rate (being D^{-1} the "residence time") and Q is the vector of the produced gas flow rates.

1.5.2 Mass and Energy Balance of the Bioprocess

A bioreactor is defined by IUPAC (1997) as an apparatus in which any bioreaction is carried out. A bioreactor is any vessel in which a chemical reaction is carried out involving biochemically active organisms (metabolic pathway) or substances derived from such organisms. Examples of these bioreactors include fermenters or water treatment plants.

Bioprocess modeling is a useful tool to accomplish several important tasks:

i. it can be the basis for adequate optimization and control technique applications
ii. it can provide necessary information about the features of the chosen bioprocessing system
iii. it synthesizes the characteristics of the specified living cells' evolution and hence, it is the best technique to predict the process efficiency.

The models show the complex biosystems attributes; so, they must be as extensive possible as and non-speculative. Moreover, the models are an acceptable compromise between the presentation of processes in detail, with considerable numbers of parameters, and the use of few parameters, easy to apply and estimate.

Most important properties of a biological mathematical model were defined in the Edwards and Wilke' postulates:

i. it is capable to represent all the culture phases
ii. it is flexible enough to approximate different data types without the insertion of significant distortions
iii. it must be continuously derivable
iv. it must be easy to operate, once the parameters are evaluated
v. each model parameter is to have a physical significance and must be easy to evaluate.

Figure 1.25 presents a simplified conceptual scheme explaining the principle of a bioreactor. It is mainly a culture vessel of volume V where the microorganisms grow. A pipe feeds the vessel with an influent medium (with flow rate F_{in}) and another one withdraws the culture medium with a flow rate F_{out}.

1.5.2.1 Dynamical Mass Balance

Component A (substrate, product, biomass)

$$\sum \begin{bmatrix} mass\ inflow \\ of\ A\ into \\ the\ reactor \\ g/d \end{bmatrix} - \sum \begin{bmatrix} mass\ outflow \\ of\ A\ from \\ the\ reactor \\ g/d \end{bmatrix} \pm \sum \begin{bmatrix} mass\ of\ A\ produced \\ /consumed\ via \\ reactions \\ g/d \end{bmatrix} = \begin{bmatrix} time\ variation \\ of\ the\ mass\ of\ A \\ g/d \end{bmatrix}$$

(1.25)

Recall that:

$$\frac{d(VA)}{dt} = F_{in}A_{in} - F_{out}A + V\mu$$

(1.26)

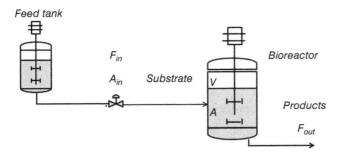

Figure 1.25 Conceptual scheme explaining the principle of a bioreactor.

Recall that:

$$\frac{d(VA)}{dt} = V\frac{dA}{dt} + A\frac{dV}{dt} \tag{1.27}$$

Therefore:

$$\frac{dA}{dt} = \frac{F_{in}}{V}A_{in} - \frac{F_{out}}{V}A - \frac{A}{V}\frac{dV}{dt} + \mu \tag{1.28}$$

1.5.2.2 Batch Process
- Significant changes of process variables over time.
- Requires more complex control.
- Requires experience with the process (feed forward control).

$$F_{in} = F_{out} = 0 \rightarrow V \; constant \;\; \frac{dV}{dt} = 0 \rightarrow \frac{dA}{dt} = \mu \tag{1.29}$$

1.5.2.3 Fed-Batch
This fed batch mode is characterized by the predetermined or controlled addition of nutrients into the bioreactor at certain times of fermenter operations. This mode of feeding nutrients into the fed batch bioreactor will allow temporal variations in the supply of nutrients. The beautiful thing about fed batch operation of a bioreactor is that it allows us a degree of control on the process and operations of the fed batch bioreactor. We can control the rate of growth of the microorganisms or the concentration of the biomass by controlling parameters such as frequency of feeds, hydraulic loadings, and concentrations of feed. The mathematical development that is going to be presented here has the following assumptions:

- The feed is provided at a constant rate.
- The production of mass of biomass per mass of substrate is constant during the fermentation time.
- A very concentrated feed is being provided to the fermentor in such a way that the change in volume is negligible.

$$F_{out} = 0 \rightarrow \frac{dV}{dt} = F_{in} \rightarrow \frac{dA}{dt} = \frac{F_{in}}{V}A_{in} - \frac{A}{V}\frac{dV}{dt} + \mu \tag{1.30}$$

1.5.2.4 Continuous
Steady state processes (chemostat)

- Constant process conditions.
- More simple process control.
- Feedback control often sufficient.

$$F_{in} = F_{out} \neq 0 \quad and \; V \; constant$$

$$\rightarrow \frac{dA}{dt} = \frac{F_{in}}{V} A_{in} - \frac{F_{in}}{V} A + \mu \tag{1.31}$$

1.5.2.5 Energy Balance
- Open system, First Law of thermodynamics

$$\Sigma \begin{bmatrix} Total \; input \\ power \; flow \end{bmatrix} - \Sigma \begin{bmatrix} Total \; output \\ power \; flow \end{bmatrix} + \Sigma \begin{bmatrix} Heat \; transfer \\ over \; the \; system \end{bmatrix}$$

$$- \Sigma \begin{bmatrix} Work \; done \\ by \; the \; system \end{bmatrix} = \begin{bmatrix} energy \\ changes \end{bmatrix} \tag{1.32}$$

Example 1.4 *Simple Microbial Growth*
Biomass growth:

$$\frac{dX}{dt} = \mu X - DX \tag{1.33}$$

Substrate consumption:

$$\frac{dS}{dt} = - \overbrace{k_1}^{yield\;coefficient} \underbrace{\overbrace{\mu}^{growth\;rate}}_{specific\;growth\;rate} X + \overbrace{D}^{dilution\;rate} S_{in} - DS \tag{1.34}$$

$$D = \frac{F_{in}}{V} \tag{1.35}$$

Example 1.5 *Heat Transfer*
Work is a form of energy, work cannot be stored. It is positive if done on the system, the work done by the system is negative.

Heat is the energy that crosses through the boundary of a system due to a temperature difference between the system and the environment. It is also known as caloric flow. To carry out a complete heat transfer analysis it is necessary to consider three different mechanisms, conduction, convection, and radiation.

Conduction, heat transfer by conduction can be performed in any of the three states of matter: liquid, gaseous, and solid, based on Fourier's law:

$$q_k = -k \frac{dT}{dx} \tag{1.36}$$

where

$q_k = [Wm^{-2}]$ Heat flow in the x-direction $k = [Wm^{-1}K^{-1}]$. The proportionality constant is a transport property known as thermal conductivity and is a characteristic of the contact material.

Example 1.6 *The Dynamics of Other Components*
Liquid product (growth associated)

$$\frac{dP}{dt} = k_2\mu X - D \quad \overbrace{P}^{\text{liquid product concentration}} \tag{1.37}$$

Gaseous product

$$\overbrace{Q}^{\text{gaseous product mass flowrate}} = k_3\mu X \tag{1.38}$$

Aerobic cultures ---> (dissolved) oxygen

$$\frac{d\quad C}{dt} = -\overbrace{k_4}^{\text{yield coefficient}}\underbrace{\overbrace{\mu}^{\text{growth rate}}}_{\text{specific growth rate}}X - DC + \underbrace{Q_{O_{2,in}} - Q_{Oout}}_{\substack{\text{gaseous oxygen} \\ \text{inlet and outlet} \\ \text{flowrates}}} \tag{1.39}$$

with $\dfrac{d\ C}{dt}$ labelled *dissolved oxygen concentration*.

If the gas-liquid transfer is limiting:

$$\underbrace{Q_{O_{2,in}} - Q_{Oout}}_{\substack{\text{gaseous oxygen} \\ \text{inlet and outlet} \\ \text{flowrates}}} = \underbrace{k_L a}_{\substack{\text{gas-liquid transfer coefficient}}}\left(\overbrace{C_s}^{\substack{\text{oxygen saturation} \\ \text{concentration}}} - C\right) \tag{1.40}$$

Coefficient of oxygen transfer.

$$k_L a_{20} = 166\left(1 - exp\left(\frac{-Q_{air}}{23040}\right)\right) \tag{1.41}$$

$$k_L a = kla_{20}1.02^{(T-20)} \quad [=]\ h^{-1} \tag{1.42}$$

Q_{air} : Airflow

Unfortunately, no (stirred-tank) bioreactor scale-down design criteria are universally applicable. Not all design parameters can be maintained as identical between large and small scales: e.g. volumetric oxygen transfer coefficient ($k_L a$), shear rate (a function of the impeller tip speed), and impeller pumping capacity (or flow). A scale-down model using a culture vessel of geometrically similar scale can be based on

- equal gassed power input per volume
- equal $k_L a$
- equal shear rate
- equal mixing time
- or a combination of oxygen transfer rate (OTR), shear rate, and mixing parameters.

1.5.3 Black Box, White Box, and Gray Box Models

1.5.3.1 Black Box

Black-box is a system in terms of inputs and outputs. One can make no assumptions about its behavior or state beyond that specified by its external operation conditions (not taking into account the chemical, physical, and microbial growth kinetics processes occurring). These models allow a global characterization of the model by disturbing its input to observe the variation or effect of said input on the states and parameters of the system at the output, that is to say: an identification, parametric sensitivity and confidence interval of operation and parameters.

Examples of some tools for the development of black-box models are

  (i) support vector machines (SVM)
 (ii) partial least squares (PLS)
(iii) artificial neural networks (ANN)
(iv) fuzzy inference system [101–104].

In the field of anaerobic digestion, ANN and PLS are used. The black-box view falls down where we must discuss aspects of a service that do not appear in the interface.

1.5.3.2 White Box

Detailed mechanistic models (white box) require:

  i. large sets of well identified parameters
 ii. large uncertainty in parameters
iii. internals being completely exposed to the user
iv. complexity of the implementation.

1.5.3.3 Gray Box

Gray boxes:

  i. should not study the source code and can lean on the results obtained by testing the user interface (possibly gathered purely by experimentation)
 ii. knowing the internal structure of the program can create a more varied and smart scenario
iii. tests cannot hope to provide a complete coverage of the program
iv. view allows only partial exposure of the system behavior.

1.5.4 Linear and Nonlinear Models

A linear model has to satisfy the superposition principle, i.e. any linear combination of model inputs (and state initial conditions) corresponds to the same linear combination of the states or the outputs. In brief, if the system functions can be represented entirely by linear equations, then the model is known as a linear model. If one or more of these functions are represented with a nonlinear equation, then the model is a nonlinear model. Actually, for more advanced applications (especially bioprocesses), many models are nonlinear as will be illustrated later.

1.5.5 Segregated and Non-segregated Models

In bioprocesses, cells are considered as discrete units. However, in a large number of models such discrete nature is not taken into account and a cell population in a culture is considered homogeneous. A segregated model considers the number of cells as a variable that describes the amount of biomass, while a non-segregated model handles terms of cell concentration as if these cells were dissolved in the culture medium.

1.5.6 Structured and Unstructured Models

The limitation of unstructured models is that they do not explicitly take into account the change in the physiological state of the cells, since they do not contain variables that measure the quality of the biomass as a function of the environment. The specific growth rate may depend on

- the substrate concentration
- the biomass concentration
- the product concentration
- the temperature
- the pH
- the dissolved oxygen concentration
- inhibitors' concentration
- the light intensity
- genetic modifications.

There are several expressions to describe microbial growth, the most widespread being the Monod equation, which relates the value of μ to the concentration of a component of the culture medium that is defective with respect to the requirements of the microorganism: the limiting substrate.

Where S is the limiting substrate concentration, it is the maximum specific growth rate, and is known as the saturation constant. Average bacteria have values close to $0.9\,h^{-1}$, yeasts $0.45\,h^{-1}$, and filamentous fungi $0.25\,h^{-1}$. In bioprocess systems, the dynamics of bioprocesses have been studied by means of several unstructured kinetic models of the Monod type, inhibition by substrate and product, Chen and Hashimoto, Contois and Levenspiel [105–109]. The presence of a toxic compound for the microorganisms is reflected in a lower rate of growth of the same. Not all microorganisms are affected in the same way by the same compounds. It is said that there are three basic types of inhibition, depending on the reversibility and kinetic parameter being affected. Through the constant "biokinetics" of the Monod equation for specific growth rate and substrate utilization, the model can be adjusted to take into account inhibitory factors (see, Tables 1.7–1.9) [105–109].

$$\mu_{max}\left[\frac{S}{k_s + S}\right] \tag{1.43}$$

Table 1.7 Kinetic models of inhibition.

Model	Parameter affected	Equation
Non-competitive inhibition	Maximum growth rate	$\mu_m \left[\dfrac{S}{k_s + S} \right] \left[\dfrac{k_I}{k_I + I} \right]$
Competitive inhibition	Saturation constant	$\mu_m \left[\dfrac{S}{K_s \left(1 + \dfrac{I}{k_I} \right) + S} \right]$
Competitive inhibition	Maximum and constant saturation rate	$\left[\dfrac{\mu_m}{1 + \dfrac{k_S}{S} + \dfrac{I}{k_I}} \right]$

Table 1.8 Kinetic models of inhibition by product (the expression of the competitive inhibition is also referred to as Haldane kinetics and has been used to express inhibition by the (*S*) or the product (*P*) itself).

Model	Equation
Boulton	$\left[\dfrac{\mu_m k_P}{P + k_P} \right]$
Levenspiel	$\mu_m \left[1 - \dfrac{P}{k_P} \right]^m$
Luong	$\mu_m \left[1 - \left(\dfrac{P}{k_P} \right)^m \right]$

1.5.7 Structured Models

Modern modeling of bioreactors emphasizes the inherent heterogeneity of distribution in large bioreactors. To model the cellular population response under these conditions, the simple structured models (intracellular) offer an alternative to the black box, unstructured models, such as Monod, in which biomass is characterized only by its quantity.

Although detailed models of the cell have been presented, the number of parameters necessary for its description leads to large experimental problems for its resolution (numerical methods, quantitative analysis.). Simple structured models have been reviewed in the past decades. They can be considered as extensions of the unstructured models in which the cell population and its relevant composition are described by up to five parameters that can be measured experimentally (e.g. compartmental models are composed of interconnected sets). In general, structured models emphasize enzymatic reactions and metabolic pathways to describe (empirically) kinetics. The modeling of the cell has recently developed from the concept of in silico systems biology. The tradition of performing experiments on a hypothesis (isolating the experimental system around a gene, an enzyme and some substrates or

Table 1.9 Modified kinetic models.

Model	Equation
Haldane–Boulton	$\mu(S,P,X) = \left[\dfrac{\mu_m S}{k_s + S + S^2 k_I^{-1}}\right]\left[\dfrac{k_P}{k_P + P}\right]X$
Haldane–Levenspiel	$\mu(S,P,X) = \left[\dfrac{\mu_m S}{k_s + S + S^2 k_I^{-1}}\right]\left[1 - \dfrac{P}{k_P}\right]^m X$
Haldane–Luong	$\mu(S,P,X) = \left[\dfrac{\mu_m S}{k_s + S + S^2 k_I^{-1}}\right]\left[1 - \left(\dfrac{P}{k_P}\right)^m\right]X$
Moser–Luong	$\mu(S,P,X) = \left[\dfrac{\mu_m S^n}{k_s + S^n}\right]\left[1 - \left(\dfrac{P}{k_P}\right)^m\right]X$
Levenspiel	$\mu(S,P,X) = \mu_m \left[\dfrac{S}{k_s + S}\right]\left[1 - \left(\dfrac{P}{P*}\right)\right]X$
Aiba	$\mu = \left[\dfrac{\mu_m S^n}{k_s + S^n}\right]\exp\left(\dfrac{-S}{k_I}\right)$
Contois	$\mu = \left[\dfrac{\mu_m S}{k_s X + S}\right]$
Chen and Hashimoto	$\mu = \left[\dfrac{\mu_m S}{k_s X_0 + (1 - k_s)S}\right]$

metabolites) has been given the impetus of new analytical and information processing techniques, which allow numerous data and behaviors to be obtained quickly. This has led to a new systematic approach that seeks to integrate the interactions between different types of intracellular molecules as well as molecules of the same type with each other.

This approach attempts to model the flow of mass, energy, and information between the following types of molecules: deoxyribonucleic and ribonucleic acids that materialize genetic information, known as genome; molecules of this same species of smaller size and that form the tRNA and rRNA, which carry genetic information, known generically as the transcriptome. A structured model representative of intracellular dynamics can be designed as complex and as efficient, according to the degree of solution and objectives [110].

Example 1.7 *Kinetic Model for Bio-Ethanol Production*

The goal of the recently emerging interest in modeling of bioprocesses is to understand and describe quantitatively the dynamics of living cells. In order to attain this purpose, new experimental procedures and modeling techniques are needed to generate and analyze relevant biological data. Kinetic modeling is an important aim in the bio-chemical reaction engineering to design, optimize, operate, and control bio-chemical reactors. The fermentation process occurs in a bioreactor and was modeled as a batch reactor. The continuity equation of mass was applied to the batch reactor to develop mathematical equations that were used to monitor the concentrations of: the rate of substrate (cellulose) utilization, the rate of biomass formation (bacteria), the rate of cellobiose, the rate of glucose and the rate of product formation (ethanol) all in g L^{-1}, the equations are [111–113].

Batch Operating Mode

Cellobiose (B) mass balance:

$$\frac{dB}{dt} = \left(\frac{\alpha_1 C}{0.947(1 + G/\alpha_2)}\right)\left[\frac{e_0}{\alpha_{10} + e_0\alpha_3 t}\right]^n - \frac{\alpha_4 B}{1 + G/\alpha_5} \qquad (1.44)$$

Glucose (G) mass balance:

$$\frac{dG}{dt} = \left(\frac{\alpha_4 B}{0.95(1 + G/\alpha_2)}\right) - \left[\frac{\alpha_6 XG}{(\alpha_7 + G)YX_{/G}} - \alpha_8 X\right] \qquad (1.45)$$

Biomass (X) mass balance:

$$\frac{dX}{dt} = \frac{\alpha_6 XG}{(\alpha_7 + G)YX_{/G}}\left(1 - \frac{X_0}{X}\right)exp^{(-\alpha_{11}t)} \qquad (1.46)$$

Ethanol (E) mass balance:

$$\frac{dE}{dt} = \frac{\alpha_9\alpha_6 XG}{(\alpha_7 + G)YX_{/G}} \qquad (1.47)$$

The Eqs. (1.44–1.46) can be used in principle to simulate the concentration changes of cellobiose, glucose, biomass mass, and ethanol with respect to time. However, the mass balance on cellulose can be expressed in terms of cellobiose, glucose, ethanol, and cell in the culture as follows:

$$C = C_0 - 0.9G - 0.947B - 0.9E/0.511 - 1.137(X - X_0) \qquad (1.48)$$

where C_0 is the initial cellulose concentration ($g l^{-1}$), X_0 is the initial cell concentration ($g l^{-1}$), the constant 0.9 is the conversion factor of a glucan unit in cellulose to glucose, 0.511 is the inverse conversion factor of glucose to ethanol, and the constant 1.137 is the conversion factor of cellulose consumed to produce yeast (g cellulose per g dry cell) assuming the molecular formula of the yeast, *Saccharomyces cerevisiae*, to be $CH_{1.74}N_{0.2}O_{0.45}$ during anaerobic fermentation of glucose [114, 115].

Assumptions of the model are that:

1) Cellulose is converted into glucose through cellobiose. Direct conversion of cellulose to glucose is negligible.
2) The effects of external and internal mass transfers on the enzyme reaction and microorganism metabolic processes can be neglected based on the assumptions that mixing was perfect.
3) The product (ethanol) formation associated with the cell growth can be represented as

$$Glucose \overset{X,G,E}{\rightarrow} Ethanol + Cell + CO_2$$

4) The enzyme deactivation is caused by the ineffective adsorption of endo-β-1,4- glucanase and exo-β-1,4-cellobiohydrolase on the solid substrate. The ineffective complex formation rate can be expressed as [116]

$$e = \left[\frac{e_0}{\alpha_{10} + e_0 k_3 t}\right]^n$$

5) The new term is added to balance the biomass based on reference [117]

$$\left(1 - \frac{X_0}{X}\right) exp^{(-a_2 t)}$$

6) This theoretical ethanol yield is similar to other experimental and industrial data [118].

$$Y_{X/G} = 0.515$$

Here α_1 specific rate constant of cellulose hydrolysis to cellobiose (l g^{-1} h), α_2 inhibitory constant of glucose to the endo-β-1,4-glucanase and exo-β-1,4-cellobiohydrolase (g l^{-1}), α_3 specific rate constant of enzyme deactivation (L (g h)$^{-1}$), α_4 specific rate constant of cellobiose hydrolysis to glucose (h^{-1}), α_5 inhibitory constant of glucose to the glycosidase (g l^{-1}), α_6 maximum specific cell growth rate constant (h^{-1}), α_7 glucose saturation constant for the microbial growth (g l^{-1}), α_8 maintenance coefficient for endogenous metabolism of the microorganisms (h^{-1}) or total number of initial enzyme concentration, α_9 product formation coefficient associated with cell growth (dimensionless), α_{10} enzyme deactivation constant (h^{-1}), α_{11} is the constant of declining substrate-enzyme reactivity (dimensionless), n is the constant (dimensionless). Batch simulation was performed for the estimation of kinetic parameters. The experimental data served to obtain the 12 parameter values of the model developed.

The maximum ethanol concentration was about 13.75 g L^{-1} at $D = 0.005$ h^{-1}, this value was slightly lower than that in the batch experiment (13.9 g l^{-1}) because, the highest ethanol concentration was dependent on the dilution rate (see Figures 1.26 and 1.27). and rapidly decreased to zero at a washout point [117].

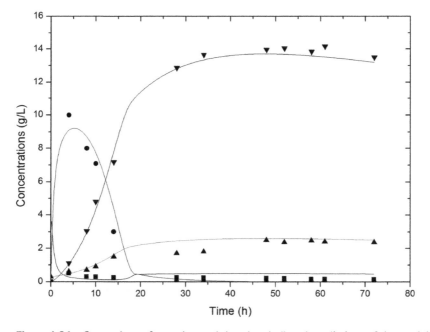

Figure 1.26 Comparison of experimental data (symbol) and predictions of the model (line).

Figure 1.27 Ethanol concentration, and productivity in continuous processes.

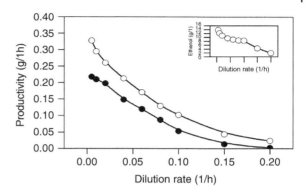

1.6 Process Optimization

1.6.1 Off-Line and On-Line Optimization of Bioprocesses

The aim of the off-line and on-line optimization method (dynamic programming) is to find the initial trajectory, that encodes the amount of substrate to be introduced into the bioreactor, in a given time interval. During the fermentation process, some of the state variables can be measured (on-line), but their values are scarcely used for closed-loop optimization purposes (nonlinear feedback control laws for special classes of nonlinear systems), and are rather employed to regulate the performance of the bioprocess. However, it is possible to develop dynamic optimization algorithms capable of quickly reacting to this new knowledge generated by updating the corresponding internal model and generating new solutions.

An alternative comes from the use of algorithms from the Evolutionary Computation (EC) field, which have been used in the past to optimize nonlinear problems with a large number of variables and parameters. These techniques have been applied with success to the optimization of feeding or real time trajectories, and, when compared with traditional methods, usually perform better [118].

However, even when the mathematical models used for open-loop optimization are reliable and validated by experimentation, in a real environment several sources of noise can contribute to changes in the observed values of the state variables. These issues are of particular importance when dealing with recombinant high-cell density fermentations, as the process, besides the nonlinearities exhibited, tends to change dramatically upon some events, like induction. Also, it is likely that there exists a time-variance of both yield and kinetic parameters not contemplated in most process models. These scenarios have an important impact on the experimental results that end up being worse than the ones predicted after running the off-line optimization [119–122]. On-line optimization algorithms generate new solutions as the bioprocesses advance, using the measurement of on-line state variables to update the proposed model, considering values of the state variables measured by sensors within the bioprocesses [118].

List of Figures

List of Tables

References

1 Mohseni, S., Vali, A.R., and Babaeipour, V. (2008). Designing appropriate schemes for the control of fed-batch cultivation of recombinant *E. coli*. *International Journal of Computers, Communications & Control* 3: 396–401.

2 de Maré, L. and Hagander, P. (2006). Parameter estimation of a model describing the oxygen dynamics in a fed-batch *E. coli* cultivation. *Biotechnology Letters* 27: 983–990.

3 Wellstead, P. (2007). Control opportunities in systems biology. *IFAC Proceedings* 40 (5): 1–18.

4 Bartocci, E. and Lió, P. (2016). Computational modeling, formal analysis, and tools for systems biology. *PLoS Computational Biology* 12 (1): e1004591.

5 Sontag, E.D. (2005). Molecular systems biology and control. *European Journal of Control* 11: 396–435.

6 Schmidt, H. and Jacobsen, E.W. (2004). Identifying feedback mechanisms behind complex cell behaviors. *IEEE Control Systems Magazine* 24: 91–102.

7 Carey, T.A., Mansell, W., and Tai, S.J. (2014). A biopsychosocial model based on negative feedback and control. *Frontiers in Human Neuroscience* 8: 94.

8 Doyle, F.J., Jovanovic, L., Seborg, D.E. et al. (2007). A tutorial on biomedical process control. *Journal of Process Control* 17: 571–630.

9 Isidori, A. (1995). *Nonlinear Systems Theory*, 3e. Berlin: Springer Verlag.

10 Kappel, F., Fink, M., and Batzel, J.J. (2007). Aspects of control of the cardiovascular-respiratory system during orthostatic stress induced by lower body negative pressure. *Mathematical Biosciences* 206: 273–308.

11 Kappel, F. and Peer, R.O. (1993). A mathematical model for fundamental regulation processes in the cardiovascular system. *Journal of Mathematical Biology* 31: 611–631.

12 Cheng-Ning, H., Fang-Ming, Y., Po-Chan, S. et al. (2007). Noise reduction of electrical impedance tomography via movable electrodes. *Instrumentation Science and Technology* 35 (5): 543–550.

13 National Research Council (1992). *Putting Biotechnology to Work: Bioprocess Engineering*. Washington, DC: The National Academies Press.

14 Wilkinson, M.H.F. (1997). Nonlinear dynamics, chaos-theory, and the "sciences of complexity": their relevance to the study of the interaction between host and microflora. In: *New Antimicrobial Strategies* (eds. P.J. Heidt, Rusch and D. VanderWaaij), 111–130). (old herborn university seminar monograph; vol. 10). Herborn-Dill: Inst Microecology & Biochem.

15 Ashoori, A., Ghods, A.H., Khaki-Sedigh, A. et al. (2008). Model predictive control of a nonlinear fed-batch fermentation process. In: *10th International Conference on Control, Automation, Robotics and Vision*, 1397–1400. Hanoi.

16 Monod, J. (1950). La technique de culture continue, théorie et applications. *Annales de l'Institut Pasteur* 79: 390–410.

17 López-Pérez, P.A., Aguilar-López, R., and Neria-González, M.I. (2015). Cadmium removal at high concentration in aqueous medium: mediated by *Desulfovibrio alaskensis*. *International Journal of Environment Science and Technology* 12: 1975–1986.

18 Kulkarni, S.J. and Kaware, D.J.P. (2013). A review on research for cadmium removal from effluent. *International Journal of Engineering Science and Innovative Technology* 2 (4): 465–469.

19 Hernández-Melchor, D.J., Cañizares-Villanueva, R.O., Terán-Toledo, J.R. et al. (2017). Hydrodynamic and mass transfer characterization of flat-panel airlift photobioreactors for the cultivation of a photosynthetic microbial consortium. *Biochemical Engineering Journal* 128: 41–148.

20 Cabrera, G., Perez, R., Gomez, J.M. et al. (2006). Toxic effects of dissolved heavy metals on *Desulfovibrio vulgaris* and *Desulfovibrio* sp. strains. *Journal of Hazardous Materials* 135: 40–44.

21 Cassidy, J., Lubberding, H.J., Esposito, G. et al. (2015). Automated biological sulphate reduction: a review on mathematical models, monitoring and bioprocess control. *FEMS Microbiology Reviews* 39 (6): 823–853.

22 López-Pérez, P.A., Neria-González, M.I., Flores-Cotera, L.B. et al. (2013). A mathematical model for cadmium removal using a sulfate reducing bacterium: *Desulfovibrio alaskensis* 6SR. *International Journal of Environmental Research* 7 (2): 501–512.

23 Goldbeter, A. (1996). *Biochemical Oscillations and Cellular Rhythms: The Molecular Bases of Periodic and Chaotic Behaviour*. Cambridge: Cambridge University Press.

24 Teusink, B., Walsh, M.C., van Dam, K. et al. (1998). The danger of metabolic pathways with turbo design. *Trends in Biochemical Sciences* 23: 162–169.

25 Iglesias, P.A. and Ingalls, B.P. (2010). *Control Theory and Systems Biology*. Massachusetts London, England: The MIT Press Cambridge.

26 Douglas, S.R. (1991). Feedback theory and Darwinian evolution. *Journal of Theoretical Biology* 152: 469–484.

27 Dimplekumar, C. and Cadet, A.C. (2014). Mathematical analysis of insulin-glucose feedback system of diabetes. *International Journal of Engineering and Applied Sciences* 5: 36–58.

28 Bernard, C., Berthault, M.F., Saulnier, C. et al. (1999). Neogenesis vs. apoptosis as main components of pancreatic beta-cell mass changes in glucose-infused normal and mildly diabetic adult rats. *FASEB Journal* 13: 1195–1205.

29 Topp, B., Promislow, K., deVries, G. et al. (2000). A model of beta-cell mass, insulin, and glucose kinetics: pathways to diabetes. *Journal of Theoretical Biology* 206: 605–619.

30 Bergman, R.N., Ider, Y.Z., Bowden, C.R. et al. (1979). Quantitative estimation of insulin sensitivity. *American Journal of Physiology* 236: 667–677.

31 Petrosyan, R.E. (2003). *Developments of the Intrinsic Low Dimensional Manifold Method and Application of the Method to a Model of the Glucose Regulatory System*. Indiana: Department of aerospace and mechanical engineering notre dame.

32 Tolic, I., Erik, M.M., and Sturis, J. (2001). Modeling the insulin–glucose feedback system: the significance of pulsatile insulin secretion. *Journal of Theoretical Biology* 207 (3): 361–375.

33 Boutayeb, W., Lamlili, M., Boutayeb, A. et al. (2014). Mathematical modelling and simulation of β-cell mass, insulin and glucose dynamics: effect of genetic predisposition to diabetes. *Journal of Biomedical Science and Engineering* 7: 330–342.

34 OECD Primer. *The Application of Biotechnology to Industrial Sustainability-a Primer*, http://www.oecd.org/sti/emerging-tech/1947629.pdf (accessed 1 July 2017).

35 Ian Marison. *Overview of Upstream and Downstream Processing of Biopharmaceuticals*, Dublin City University http://www.engineersirelandcork.ie/downloads/Biopharmaceuticals%2020Jan09%20-%202%20-%20Ian%20Marison%20DCU.pdf *Bioprocess Engineering and Head of School of Biotechnology*, (accessed 1 July 2017).

36 Alford, J.S. (2006). Bioprocess control: advances and challenges. *Computers and Chemical Engineering* 30: 1464–1475.

37 Gronemeyer, P., Ditz, R., and Strube, J. (2014). Trends in upstream and downstream process development for antibody manufacturing. *Bioengineering* 1 (4): 188–212.

38 Butler, M. and Meneses-Acosta, A. (2012). Recent advances in technology supporting biopharmaceutical production from mammalian cells. *Applied Microbiology and Biotechnology* 96: 885–894.

39 Li, F., Vijayasankaran, N., Shen, A. et al. (2010). Cell culture processes for monoclonal antibody production. *Pharmaceutical Sciences Encyclopedia* 2: 466–479.

40 Costa Rita, A., Rodrigues, E.M., Henriques, M. et al. (2010). Guidelines to cell engineering for monoclonal antibody production. *European Journal of Pharmaceutics and Biopharmaceutics* 74: 127–138.

41 Lee, C.J., Seth, G., Tsukuda, J. et al. (2009). A clone screening method using mRNA levels to determine specific productivity and product quality for monoclonal antibodies. *Biotechnology and Bioengineering* 102: 1107–1118.

42 Le, H., Kabbur, S., and Pollastrini, L. (2012). Multivariate analysis of cell culture bioprocess data--lactate consumption as process indicator. *Journal of Biotechnology* 162: 210–223.

43 Hernández Melchor, D. J., et al. (2016) *The potential of microalgae isolated from seawater for biodiesel production and kinetic modeling*. International-Mexican Congress on Chemical Reaction Engineering (IMCCRE 2016) June 2016.

44 Kumar, M.A. (2011). *Biosensors and Automation for Bioprocess Monitoring and Control*. Lund University (MediaTryck).

45 Richards, M.A., Cassen, V., Heavner, B.D. et al. (2014). Media DB: a database of microbial growth conditions in defined media. *PLoS One* 9 (8): e103548.

46 Keasling, J.D. (2011). Manufacturing molecules through metabolic engineering. In: *Institute of Medicine (US) Forum on Microbial Threats*. The Science and Applications of Synthetic and Systems Biology: Workshop Summary, A10. Washington (DC): National Academies Press (US) Available from: https://www.ncbi.nlm.nih.gov/books/NBK84444.

47 Kumar, A.J., Mahalakshmi, D.K., and Sridhari, G. (2011). Integrated application of bioprocess engineering and biotechniques for quality & bulk drug manufacturing. *Journal of Bioequivalence and Bioavailability* 3: 277–285.

48 Anesiadis, N. (2014). *A Synthetic Biology Application in Metabolic Engineering*. Canadá: University of Toronto URI http://hdl.handle.net/1807/68110 (accessed 1 July 2017).

49 Doran, P.M. (2013). *Bioprocess Engineering Principles*, 2e, 899–919. Academic Press.

50 Farías-Álvarez, L.J., Gschaedler-Mathis, A., Sánchez-Ortiz, A.F. et al. (2018). Xanthophyllomyces dendrorhous physiological stages determination using combined measurements from dielectric and Raman spectroscopies, a cell counter system and fluorescence flow cytometry. *Biochemical Engineering Journal* 136: 1–8.

51 Julien, C. and Whitford, W. (2007). Bioreactor monitoring, modeling, and simulation. *Bioprocess International*: 10–17.

52 Salehi-Nik, N., Amoabediny, G., Pouran, B. et al. (2013). Engineering parameters in bioreactor's design: a critical aspect in tissue engineering. *BioMed Research International*: 762132.

53 Locher, G., Sonnleitner, B., and Fiechter, A. (1992). On-line measurement in biotechnology: techniques. *Journal of Biotechnology* 25: 23–53.

54 Chuang, G.S., Ho, P.Y., and Li, H.Y. (2004). *The Determination of Multiple Steady States in Two Families of Biological Systems*, vol. 59a, 136–146. Mainz: Verlag der Zeitschrift für Naturforschung Tübingen.

55 Calderón-Soto, L.F., Herrera-López, E.J., Lara-Cisneros, G. et al. (2019). On unified stability for a class of chemostat model with generic growth rate functions: maximum yield as control goal. *Journal of Process Control* 77: 61–75.

56 Caramihai, M. and Severin, I. (2013). *Bioprocess Modeling and Control, Biomass Now – Sustainable Growth and Use* (ed. M.D. Matovic). InTech Available from: https://www.intechopen.com/books/biomass-now-sustainable-growth-and-use/bioprocess-modeling-and-control.

57 Rajinikanth, V. and Latha, K. (2012). Modeling, analysis, and intelligent controller tuning for a bioreactor: a simulation study. *ISRN Chemical Engineering* 413657: 15.

58 Sivakumaran, N., Radhakrishnan, T.K., and Babu, J.S.C. (2006). Identification and control of bioreactor using recurrent networks. *Instrumentation Science and Technology* 34: 635–651.

59 Kumar, M., Giriraj, J.R., Anantharaman, N. et al. (2008). Genetic algorithm based PID controller tuning for a model bioreactor. *Institute of Chemical Engineers* 50: 214–226.

60 Foss, B.A., Johansen, T.A., and Sorensen, A.V. (1995). Nonlinear predictive control using local models-applied to a batch fermentation process. *Control Engineering Practice* 3: 389–396.

61 Oliveira, R. (2004). Combining first principles modeling and artificial neural networks: general framework. *Computers and Chemical Engineering* 28: 755–766.

62 Renard, F., Vande Wouwer, A., and Valentinotti, S. (2006). A practical robust control scheme for yeast fed-batch cultures- an experimental validation. *Journal of Process Control* 16: 855–864.

63 López-Pérez, P., Neria-González, M., and Aguilar-López, R. (2015). Improvement of activated sludge process using a nonlinear PI controller design via adaptive gain. *International Journal of Chemical Reactor Engineering* 14 (1): 407–416.

64 Horiuchi, J.I. (2002). Review: fuzzy modeling and control of biological Processes. *Journal of Bioscience and Bioengineering* 94: 574–578.

65 Gomes, J. and Menawat, A.S. (1998). Precise control of dissolved oxygen in bioreactors- a model-based geometric algorithm. *Chemical Engineering Science* 55: 67–78.

66 Samorski, M., Müller-Newen, G., and Büchs, J. (2005). Quasi-continuous combined scattered light and fluorescence measurements: a novel measurement technique for shaken microtiter plates. *Biotechnology and Bioengineering* 92: 61–68.

67 Quiroz, G., Flores-Gutiérrez, C.P., and Femat, R. (2011). Suboptimal H-infinity hyperglycemia control on T1DM accounting biosignals of exercise and nocturnal hypoglycemia. *Optimal Control Applications & Methods* 32: 239–252.

68 Quiroz, G. and Femat, R. (2010). Theoretical blood glucose control in hyper-hypoglycemic and exercise scenarios by means of a $H-infinity decision algorithm. *Journal of Theoretical Biology* 263: 154–160.

69 Femat, R., Ruiz-Velázquez, E., and Quiroz, G. (2009). Weighting restriction for intravenous insulin delivery on T1DM patient via H-infinity control. *IEEE Transactions on Automation Science and Engineering* 6: 239–247.

70 Hernández-Medina, A., Flores-Gutiérrez, C.P., and Femat, R. Robustness properties preservation in sub-optimal T1DM H-inf. Control: w-SPR substitutions. *Optimal Control Applications & Methods* 39: 220–229.

71 Leyva, H., Carrillo, F., Quiroz, A.G. et al. (2016). Robust stabilization of positive linear systems via sliding positive control. *Journal of Process Control* 41: 47–55.

72 Leyva, H., Carrillo, F., Quiroz, A.G. et al. (2019). Insulin stabilisation in artificial pancreas: a positive control approach. *IET Control Theory & Applications* 13: 970–978.

73 Aguilar-López, R., Lara-Cisneros, G., and Femat, R. (2017). Dynamic nonlinear feedback control applied to improve butanol production by Clostridium acetobutylicum. *International Journal of Chemical Reactor Engineering* 34: 1–17.

74 Lara-Cisneros, G., Femat, R., and Dochain, D. (2017). Robust sliding mode based extremum-seeking controller for reaction systems via uncertainty estimation approach. *International Journal of Robust and Nonlinear Control* 27: 3218–3235.

75 Lara-Cisneros, G., Aguilar, R., Dochain, D., and Femat, R. (2016). On-line estimation of VFA concentration in anaerobic digestion via methane outflow rate measurements. *Computers & Chemical Engineering* 94: 250–256.

76 Lara-Cisneros, G., Alvarez-Ramírez, J., and Femat, R. (2016). Self-optimizing control for a class of continuous bioreactor via variable-structure feedback. *International Journal of Systems Science* 47 (6): 1394–1406.

77 Lara-Cisneros, G., Aguilar-López, R., and Femat, R. (2015). On the dynamic optimization of methane production in anaerobic digestion via extremum-seeking control approach. *Computers & Chemical Engineering* 75: 49–59.

78 Lara-Cisneros, G., Femat, R., and Dochain, D. (2014). An extremum seeking approach via variable-structure control for fed-batch bioreactors with uncertain growth rate. *Journal of Process Control* 24: 663–671.

79 Rodríguez, G., Quiroz, R. Femat, et al. (2015). An adaptive observer for operation monitoring of anaerobic digestion wastewater treatment. *Chemical Engineering Journal* 269: 186–193.

80 Rocchitta, G., Spanu, A., Babudieri, S. et al. (2016). Enzyme biosensors for biomedical applications: strategies for safeguarding analytical performances in biological fluids. *Sensors* 16 (6): 780.

81 Gnoth, S., Jenzsch, M., and Simutis, R. (2008). Control of cultivation processes for recombinant protein production: a review. *Bioprocess and Biosystems Engineering* 31: 21–39.

82 Faassen, S.M. and Hitzmann, B. (2015). Fluorescence spectroscopy and chemometric modeling for bioprocess monitoring. *Sensors* 15 (5): 10271–10291.

83 Tohmola, N., Ahtinen, J., Pitkänen, J.P. et al. (2011). On-line high performance liquid chromatography measurements of extracellular metabolites in an aerobic batch yeast (*Saccharomyces cerevisiae*) culture. *Biotechnology and Bioprocess Engineering* 16: 264–272.

84 Munisamy, S.M., Chambliss, C.K., and Becker, C.S. (2012). Direct infusion electrospray ionization-ion mobility high resolution mass spectrometry (DIESI-IM-HRMS) for rapid characterization of potential bioprocess streams. *Journal of the American Society for Mass Spectrometry* 23 (7): 1250–1259.

85 Chung, S., Wen, X., Vilholm, K. et al. (1991). Michael De Bang, Gary Christian, Jaromir Ruzicka, Novel flow-injection analysis method for bioprocess monitoring. *Analytica Chimica Acta* 249: 77–85.

86 Hansen, K.F., Lauritsen, F.R., and Degn, H. (1994). An on-line sampling system for fermentation monitoring using membrane inlet mass spectrometry (MIMS): application to phenoxyacetic acid monitoring in penicillin fermentation. *Biotechnology and Bioengineering* 44: 347–353.

87 Turkia, H., Holmström, S., Paasikallio, T. et al. (2013). Online capillary electrophoresis for monitoring carboxylic acid production by yeast during bioreactor cultivations. *Analytical Chemistry* 85: 9705–9712.

88 Wu, M.J., Ye, G.F., Wang, C.H. et al. (2017). The use of a gas chromatography/Milli-whistle technique for the on-line monitoring of ethanol production using microtube array membrane immobilized yeast cells. *Analytical Science* 33 (5): 625–630.

89 Pitkänen, J.P. and Turkia, H. *Literature Review of On-line Bioprocess* seventh framework programme theme 2 food, agriculture and fisheries, and biotechnology nano- and microtechnology-based analytical devices for online measurements of bioprocesses Deliverable D.8.3 GA no. 227243 (accessed 1 July 2017). https://www.vtt.fi/files/sites/nanobe/literature_review.pdf

90 Warth, B., Rajkai, G., and Mandenius, C.F. (2010). Evaluation of software sensors for on-line estimation of culture conditions in an *Escherichia coli* cultivation expressing a recombinant protein. *Journal of Biotechnology* 147: 37–45.

91 Soons, Z.I.T.A., Voogt, J.A., van Straten, G. et al. (2006). Constant specific growth rate in fed-batch cultivation of *Bordetella pertussis* using adaptive control. *Journal of Biotechnology* 125: 252–268.

92 Nor, Z.M., Tamer, M.I., Scharer, J.M. et al. (2001). Automated fed-batch culture of *Kluyveromyces fragilis* based on a novel method for on-line estimation of cell specific growth rate. *Biochemical Engineering Journal* 9: 221–231.

93 Cannizzaro, C., Valentinotti, S., and von Stockar, U. (2004). Control of yeast fed-batch process through regulation of extracellular ethanol concentration. *Bioprocess and Biosystems Engineering* 26: 377–383.

94 Biener, R., Steinkaemper, A., and Hofmann, J. (2010). Calorimetric control for high cell density cultivation of a recombinant *Escherichia coli* strain. *Journal of Biotechnology* 146 (1–2): 45–53.

95 Venkateswarlu, C. (2004). Advances in monitoring and state estimation of bioreactors. *Journal of Scientific and Industrial Research* 63 (6): 491–498.

96 Rocha, I., Veloso, A., Carneiro, S. et al. (2008). *Implementation of a specific rate controller in a fed-batch E. coli fermentation*. In: *Proceeding of the 17th IFAC World Congress*, 15565–15570. Seoul, Korea: International Federation of Automatic Control (IFAC).

97 Kadlec, P., Gabrysa, B., and Strandt, S. (2009). Data-driven soft sensors in the process industry. *Computers and Chemical Engineering* 33: 795–814.

98 López Pérez, P.A., Neria González, M.I., and Aguilar-Lopez, R. (2016). Concentrations monitoring via software sensor for bioreactors under model parametric uncertainty: application to cadmium removal in an anaerobic process. *Alexandria Engineering Journal* 55 (2): 1893–1902.

99 Zomorrodi, A.R. and Segrè, D. (2016). Synthetic ecology of microbes: mathematical models and applications. *Journal of Molecular Biology* 428: 837–861.

100 Yur'evna, R.G. (2000). *Mathematical Models in Biophysics*. Lomonosov Moscow State University.

101 Desai, K., Badhe, Y., Tambe, S.S. et al. (2006). Soft-sensor development for fed-batch bioreactors using support vector regression. *Biochemical Engineering Journal* 27: 225–239.

102 Nandi, S., Ghosh, S., Tambe, S.S. et al. (2001). Artificial neural network assisted stochastic process optimization strategies. *AICHE Journal* 47: 126–141.

103 Dohnal, M. (1985). Fuzzy bioengineering models. *Biotechnology and Bioengineering* 27: 1146–1151.

104 Koch, C., Posch, A.E., Goicoechea, H.C. et al. (2014). *Multi-analyte quantification in bioprocesses by Fourier-transform-infrared spectroscopy by partial least squares regression and multivariate curve resolution. Analytica Chimica Acta* 807 (100): 103–110.

105 Beyenal, H., Chen, Suet, N., and Lewandowski, Z. (2003). The double substrate growth kinetics of *Pseudomonas aeruginosa. Enzyme and Microbial Technology* 32: 92–98.

106 Moser, H. (1957). *The dynamics of bacterial populations maintained in the chemostat. Cold Spring Harbor Symposia on Quantitative Biology* 22: 121–137.

107 Boulton, R. (1980). The prediction of fermentation behavior by a kinetic model. *American Journal of Enology and Viticulture* 31: 40–45.

108 Levenspiel, O. (1980). The monod equation: a revisit and a generalization to product inhibition situations. *Biotechnology and Bioengineering* 22: 1671–1687.

109 Luong, J.H.T. (1985). Kinetics of ethanol inhibition in alcohol fermentation. *Biotechnology and Bioengineering* 27: 280–285.

110 Edwards, J.S., Covert, M., and Palsson, B. (2002). Metabolic modelling of microbes: the flux-balance approach. *Environmental Microbiology* 4 (3): 133–140.

111 Lara-Cisneros, G., Femat, R., and Pérez, E.A. (2012). On dynamical behaviour of two-dimensional biological reactors. *International Journal of Systems Science. Principles and Applications of Systems and Integration* 43: 526–534.

112 Shen, J. and Agblevor, F.A. (2010). The operable modeling of simultaneous saccharification and fermentation of ethanol production from cellulose. *Applied Biochemistry and Biotechnology* 160 (3): 665–681.

113 Shen, J. and Agblevor, F.A. (2008). Kinetics of enzymatic hydrolysis of steam exploded cotton gin waste. *Chemical Engineering Communications* 195 (9): 1107–1121.

114 South, C.R.D., Hogsett, A.L., and Lynd, L.R. (1995). Modeling simultaneous saccharification and fermentation of lignocellulose to ethanol in batch and continuous reactors. *Enzyme and Microbial Technology* 17: 797–803.

115 Shuler, M.L. and Kargi, F. (2002). *Bioprocess Engineering*, 2e. Prentice Hall PTR, 217.

116 Kadam, L.K., Rydholm, E.C., and McMillan, J.D. (2004). Development and validation of a kinetic model for enzymatic saccharification of lignocellulosic biomass. *Biotechnology Progress* 20: 698–705.

117 Roubos, J.A., vanStraten, G., and van Boxtel, A.J. (1999). An evolutionary strategy for fed-batch bioreactor optimization: concepts and performance. *Journal of Biotechnology* 67: 173–187.

118 Rocha, M., Pinto, J.P., Rocha, I. et al. (2007). *Evaluating evolutionary algorithms and differential evolution for the online optimization of fermentation processes*. In: *Evolutionary Computation, Machine Learning and Data Mining in Bioinformatics*. EvoBIO 2007. Lecture Notes in Computer Science, vol. 4447 (eds. E. Marchiori, J.H. Moore and J.C. Rajapakse). Berlin, Heidelberg: Springer.

119 Moriyama, H. and Shimizu, K. (1996). On-line optimization of culture temperature for ethanol fermentation using a genetic algorithm. *Journal Chemical Technology Biotechnology* 66: 217–222.

120 Ronen, M., Shabtai, Y., and Guterman, H. (2002). Optimization of feeding profile for a fed-batch bioreactor by an evolutionary algorithm. *Journal of Biotechnology* 28: 253–263.

121 López-Pérez, P.A., Puebla, H., Velázquez Sánchez, H.I. et al. (2016). Comparison tools for parametric identification of kinetic model for ethanol production using evolutionary optimization approach. *International Journal of Chemical Reactor Engineering* 14 (6).

122 Nickel, D.B., Cruz-Bournazou, Wilms, M.N. et al. (2017). On line bioprocess data generation, analysis, and optimization for parallel fed-batch fermentations in milliliter scale. *Engineering in Life Sciences* 17: 1195–1201.

2

Mathematical Preliminaries

In this chapter, the mathematical framework necessary for the analysis and characterization of bioprocess dynamics is established. In particular, the fundamental concepts and
tools from the theory of the stability of dynamical systems is explained and an overview
of the non-smooth dynamical systems is given In other words, the chapter deals with the
mathematical approach we follow to describe the evolution in time of the bioprocess under
consideration. Therefore, understanding and formalizing the role of nonlinearity is indeed
one of the greatest challenges in the study of living systems, mainly in bioprocesses. Design,
the startup of high-performance control and monitoring of the bio-systems is essentially a
problem of obtaining a nominal (nonlinear) model for the dynamic characteristics of the
bio-system considered for deriving a control law that achieves the desired objective using
the nominal model. However, should the changes become large, the controller, as originally
designed with fixed parameters, will fail to meet the design specifications, e.g. stability,
chaos, state, output tracking, or state regulation. In engineering practice two of the most
important sources of modeling error are the presence of nonlinearities in the system and
lack of exact knowledge of some of the system parameters, consequently, it is necessary to
describe the properties of the state of the dynamics of the nominal model under the theory
of systems.

2.1 Systems of Ordinary Differential Equations

Ordinary differential equation (ODE) allow us to study all kinds of evolutionary processes
with the properties of determinacy, differentiability, and finite-dimensionally. The ODE
study began very soon after the invention of differential and integral calculus. In 1671, Isaac
Newton had laid the foundations for the study of differential equations. He was followed
by Leibnitz who named differential equation in 1676 to denote the relationship between
different dx and dy of two variables x to y (infinitesimal distance) [1].

More precisely, an equation that involves derivatives of one or more unknown dependent variables with respect to one or more independent variables is called a differential

Control in Bioprocessing: Modeling, Estimation and the Use of Soft Sensors, First Edition.
Pablo Antonio López Pérez, Ricardo Aguilar López, and Ricardo Femat.
© 2020 John Wiley & Sons Ltd. Published 2020 by John Wiley & Sons Ltd.

equation. Also, when a differential equation involves one or more derivatives with respect to a particular variable, that variable is called the independent variable. A variable is called dependent if a derivative of that variable occurs. In the beginning, mathematicians were mostly engaged in formulating differential equations and solving them, they did not worry about the existence and uniqueness of solutions [1].

An equation involving ordinary derivatives of one or more dependent variables with respect to a single independent variable is called an ODE. A general form of an ODE containing one independent and one dependent variable is $f(x, y, y', y'', \ldots, y^n) = 0$ where f is an arbitrary function of $x, y, y', \ldots, y^n$, here x is the independent variable while y is the dependent variable and $y^n = \frac{d^n y}{dx^n}$. The order of an ODE is the order n of the highest derivate appearing in it. On the other hand, partial differential equations are those which have two or more independent variables. For example, consider the following differential equation

$$\dot{y} + a(x)y = b(x) \} \, a(x), b(x) \text{ are scalar functions and } x \in \mathfrak{R}^n. \tag{2.1}$$

Observations about Eq. (2.1)

- $\dot{y}$ is the derivative of y with respect to x
- is first order because the highest derivative that appears in it is a first order derivative
- if $b(x)$ is zero, then (2.1) is homogeneous, otherwise, it is non-homogeneous.

2.1.1 Differential Equations, Vector Fields, and State-Space Description

A common class of mathematical models for dynamical systems are ODEs. Mathematically, an ODE is written as.

$$\frac{dx}{dt} = f(x) \tag{2.2a}$$

Here $x = (x_1, x_2, \ldots, x_n) \in \mathfrak{R}^n$ is a vector of real numbers that describes the current state of the system and Eq. (2.2a) describes the rate of change of the state as a function of the state itself. Note that we do not bother to write the vector x any differently than a scalar variable. It will generally be clear from the context whether a variable is a vector or scalar quantity [2]. The function $f: \mathfrak{R}^n \rightarrow \mathfrak{R}^n$ is smooth.

The state of a system is the minimal set $\{x_i(t), \forall i = 1, 2, 3, \ldots, n\}$, needed together with the input $u(t)$ with t in the interval $[t_0, t_i)$ to uniquely determine the behavior of the system in the interval $[t_0, t_i]$. The index n is known as the order of the system. As time t increases, the state of the system evolves and each $x_i(t)$ becomes a time variable. Such variables are known as the state variables. In *vector* notation, the set of state variables form the state vector

The evolution of the states can be described using either a time plot or a phase plot, both of which are shown in Figures 2.1 and 2.2. The model is applied to a wastewater treatment plant. The time plot, Figure 2.2, shows the values of the individual states as a function of time. The phase plot, Figure 2.3, shows the vector field for the system, which gives the state velocity at every point in the state space. In addition; we have superimposed the traces of some of the states from different conditions (represented as an arrow).

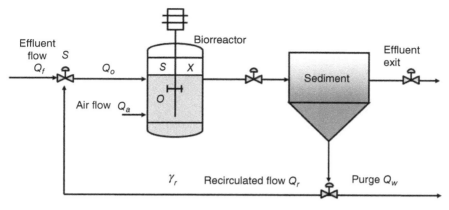

Figure 2.1 Schematic diagram of the activated-sludge process.

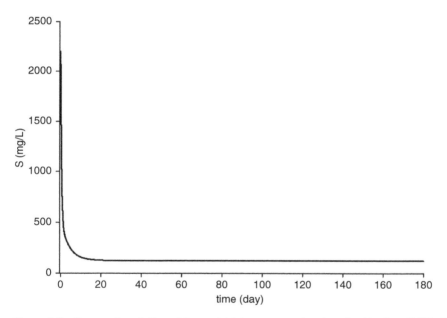

Figure 2.2 Temporal evolution of the variable's concentration described by Eqs. (2.2b)–(2.2d).

Example 2.1 *Mathematical Model of an Activated Sludge Process*

A description of the wastewater plant, basic activated sludge process has several interrelated steps. These modules are (see Figure 2.1):

- aeration systems,
- aeration source,
- clarifier-sediment,
- recycle, waste.

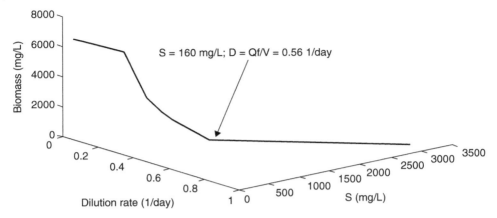

Figure 2.3 Phase diagram of the open-loop system of output variables, the effect of different dilution rates.

The mathematical model considered here is presented in [3, 4]. The wastewater flow produced is about 7000 m³ d⁻¹ and contains volatile organic carbon substances classified as toxics like 1,2 dichloroethane, chloroform, benzene, among other volatile compounds, (VOC's).

The model presented in [5], it is based on activated sludge model no 1 (ASM1), considering (2.2a) and a carbon removal model with a dynamic energy balance to introduce the temperature effects on the maximum specific growth rate, mass transfer coefficient for oxygen (kla) and death coefficient (kd), which were incorporated in the mass balance equations of the process.

In the bioreactor:

Substrate (S) chemical oxygen demand (COD) concentration mass balance:

$$\underbrace{\frac{dS}{dt}}_{\text{independent variable}}\overset{\text{dependent variable}}{} = \frac{Q_f}{V}S_f - \frac{Q_0}{V}S - \overset{\text{Parameters}}{\frac{\mu_{max}}{Y}}\left(\frac{S}{K_s + S}\right)\left(\frac{O}{K_{OH} + O}\right)X$$

$$+ \underbrace{k_d(1 - f_n)X}_{\text{Parameters}} - \underbrace{k_{ev}}_{\text{Parameters}} S \tag{2.2b}$$

(X) concentration balance:

$$\frac{dX}{dt} = \frac{Q_r}{V}Y_r - \frac{Q_0}{V}X + \frac{\mu_{max}}{Y}\left(\frac{S}{K_s + S}\right)\left(\frac{O}{K_{OH} + O}\right)X - k_dX \tag{2.2c}$$

Oxygen (O) concentration mass balance:

$$\frac{dO}{dt} = \frac{Q_f}{V}O_f - \frac{Q_0}{V}O - \frac{\mu_{max}}{Y_{O_2}}\left(\frac{Y_s}{K_s + Y_s}\right)\left(\frac{O}{K_{OH} + O}\right)X - \overset{\text{Parameters}}{k_La}(O_{sat} - O) \tag{2.2d}$$

The phase portrait provides a valuable intuitive representation of the equation as a vector field or a flow. While systems of second order (two states) can be represented in this way, complicated abstractions are required to visualize equations and state space of higher order using this approach. If the state of a natural system is known at some time, its future development is completely determined. One of the applications of vector fields is the visualization of the solution of ODEs. Differential equations define a vector field at every point in the phase space. The solution of a differential equation with prescribed initial condition follows the flow of vectors [6].

The ODE (2.2a) is called an autonomous system because there are no external influences. In many examples, it is useful to model the effects of external disturbances or controlled forces on the system. One way to capture this is to replace Eq. (2.2a) by

$$\frac{dx}{dt} = f(x, u) \tag{2.-2}$$

where u represents the effect of external influences. The model (2.-2) is called a forced or controlled differential equation. The model implies that the rate of change of the state can be influenced by the input $u(t)$.

Adding the input makes the model richer and allows new questions to be posed. For example, we can examine what influence external disturbances have on the trajectories of a system (plant: bioprocess). Or, in the case when the input variable is something that can be regulated in a controlled way, we can analyze whether it is possible to "path-follow" the system from one point in the state space (state variable) to another through proper choice of the input (control law) [4].

The state of the dynamic system at time t is uniquely determined by the state at time t_0, the input $t \geq t_0$ and is independent of the state and input before t_0 (this property is called the "Markov Property"). In addition, the state variables of a dynamic system are the variables making up the smallest set of variables determining the state of the dynamic system. If n state variables are needed to completely describe the behavior of a given system, then those state variables can be considered the n components of a vector x called a state vector. The n-dimensional space whose coordinate axes consist of the $x_1 - axis, x_{21} - axis, \ldots, x_n - axis$ is called a state space. Large classes of engineering, biological, social, and economic systems may be represented by state-determined system models. Furthermore, the values of the system state variables at any time provide sufficient information to determine the values of all other variables in the system at that time. There is no unique set of state variables that describe any given system; many different sets of variables may be selected to yield a complete system description. However, for a given system the order n is unique and is independent of the particular set of state variables chosen. State variable descriptions of systems may be formulated in terms of physical and measurable variables, or in terms of variables that are not directly measurable. The important point is that any set of state variables must provide a complete description of the system [5, 7].

The so-called state-space description provides the dynamics as a set of coupled first-order differential equations in a set of internal variables known as state variables, together with a set of algebraic equations that combine the state variables into physical output variables. In the state-space analysis, we are concerned with three types of variables that are involved in the modeling of a dynamic system: input, output, and state variables. A description of systems that are linear and time-invariant (LTI) is observed in Eqs. (2.4) and (2.5), that is systems described by linear differential equations with constant coefficients [8]. If a system is LTI and if it is described by n state variables, r input variables, and m output variables,

Eq. (2.4), then the state equation will have the form:

$$
\begin{aligned}
\dot{x}_1 &= a_{11}x_1 + a_{12}x_2 + \ldots + a_{1n}x_n + b_{11}u_1 + b_{12}u_2 + \ldots + b_{1r}u_r \\
\dot{x}_2 &= a_{21}x_1 + a_{22}x_2 + \ldots + a_{2n}x_n + b_{21}u_1 + b_{22}u_2 + \ldots + b_{2r}u_r \\
&\vdots \quad \vdots \quad \vdots \quad \ldots \quad \vdots \quad \vdots \quad \vdots \quad \ldots \quad \vdots \\
\dot{x}_n &= a_{n1}x_1 + a_{n2}x_2 + \ldots + a_{nn}x_n + b_{n1}u_1 + b_{n2}u_2 + \ldots + b_{nr}u_r
\end{aligned}
\tag{2.4}
$$

An important property of the linear state equation description is that all system variables may be represented by a linear combination of the state variables and the system inputs, the output equation will then have the form:

$$
\begin{aligned}
y_1 &= c_{11}x_1 + c_{12}x_2 + \ldots + c_{1n}x_n + d_{11}u_1 + d_{12}u_2 + \ldots + d_{1r}u_r \\
y_2 &= c_{21}x_1 + c_{22}x_2 + \ldots + c_{2n}x_n + d_{21}u_1 + d_{22}u_2 + \ldots + d_{2r}u_r \\
&\vdots \quad \vdots \quad \vdots \quad \ldots \quad \vdots \quad \vdots \quad \vdots \quad \ldots \quad \vdots \\
y_m &= c_{m1}x_1 + a_{m2}x_2 + \ldots + a_{mn}x_n + b_{m1}u_1 + b_{m2}u_2 + \ldots + b_{mr}u_r
\end{aligned}
\tag{2.5}
$$

where the coefficients a_{ij}, b_{ij}, c_{ij} and d_{ij} are constants. If we use vector–matrix expressions, these equations can be written as:

State equation

$$
\dot{x} = Ax + Bu
\tag{2.6}
$$

$$
x = \begin{bmatrix} x_1 \\ x_2 \\ \vdots \\ x_n \end{bmatrix}, A = \begin{bmatrix} a_{11} & a_{12} & \cdots & a_{1n} \\ a_{21} & a_{22} & \cdots & a_{2n} \\ \vdots & \vdots & & \vdots \\ a_{n1} & a_{n2} & \cdots & a_{nn} \end{bmatrix}, B = \begin{bmatrix} b_{11} & b_{12} & \cdots & b_{1r} \\ b_{21} & b_{22} & \cdots & b_{2r} \\ \vdots & \vdots & & \vdots \\ b_{n1} & b_{n2} & \cdots & b_{nr} \end{bmatrix}, u = \begin{bmatrix} u_1 \\ u_2 \\ \vdots \\ u_r \end{bmatrix}
\tag{2.7}
$$

Dynamical relation, Output equation

$$
y = Cx + Du
\tag{2.8}
$$

$$
y = \begin{bmatrix} y_1 \\ y_2 \\ \vdots \\ y_n \end{bmatrix}, C = \begin{bmatrix} c_{11} & c_{12} & \cdots & c_{1n} \\ c_{21} & c_{22} & \cdots & c_{2n} \\ \vdots & \vdots & & \vdots \\ c_{m1} & c_{m2} & \cdots & c_{mn} \end{bmatrix}, D = \begin{bmatrix} d_{11} & d_{12} & \cdots & d_{1r} \\ d_{21} & d_{22} & \cdots & d_{2r} \\ \vdots & \vdots & & \vdots \\ d_{m1} & b_{m2} & \cdots & d_{mr} \end{bmatrix}
\tag{2.9}
$$

It should be noted that:

i. Matrices A, B, C and D are called the state matrix, input matrix, output matrix, and direct transmission matrix, respectively.

ii. Vectors x, u, and y are the state vector, input vector, and output vector respectively.

iii. The elements of the state vector are the state variables.

We can then rewrite n first-order differential equations as a single n-dimensional first-order vector differential equation

$$
\dot{x} = f(x, t, u)
\tag{2.10}
$$

When q measurements are present, then we include another q-dimensional equation

$$
y = h(x, t, u)
\tag{2.11}
$$

Where y represents the q measured process outputs. In summary, we refer to (2.10) as the state equation and (2.11) as the output equation. When the dynamical system does not have any control input, i.e. it runs autonomously, we refer to it as an autonomous system. Mathematically, this is written,

$$\dot{x} = f(x, t) \tag{2.12}$$

An important concept in dynamical systems is equilibrium. Roughly, the equilibrium of a dynamical system is the point in the state-space where the states remain constant, i.e. steady-state. It is defined mathematically as follows (equilibrium):

$$f(x^{eq}, t) = 0, \quad x^{eq} : \quad \text{State Equilibrium (continuous systems at steady state)} \tag{2.13}$$

Remarks

i. Systematic analysis and synthesis of higher order systems without truncation of system dynamics.
ii. Convenient tool for MIMO systems.
iii. Uniform platform for representing time-invariant systems, time-varying systems, linear systems as well as nonlinear systems.
iv. Can describe the dynamics in almost all systems (mechanical systems, electrical systems, biological systems, economic systems, social systems, etc.).

Example 2.2   *Mass-Spring-Damper Model Written on the State-Space Form*
The mass-spring-damper (MSD) system as an example, so here is a brief description of the typical MSD system in state space. The differential equation that describes an MSD is:

$$m\ddot{x} + c\dot{x} + kx = p \tag{2.14}$$

m: mass, kg
c: viscous damping coefficient, $N\,s\,m^{-1}$
x: the position of the mass, m
k: spring constant, $N\,m^{-1}$
p: force input, N

Depending on the values of m, c, and k, the system can be underdamped, overdamped or critically damped. Representing a system in state space leads to a set of first-order differential equations instead of having higher order differential equations. Having a set of first-order ODEs allows us to arrange the equations into matrix form (state space) which is a very convenient representation that can be used for simulation using numerical methods [1].
Consider

$$x_1 = x$$

$$x_2 = \dot{x} = \dot{x}_1 \tag{2.15}$$

$$\dot{x}_1 = x_2 \tag{2.16}$$

$$\dot{x}_1 = \dot{x}_2 \tag{2.17}$$

Consider (2.14)–(2.17)

$$m\ddot{x}_2 + c\dot{x}_2 + kx_1 = p \quad \therefore \tag{2.18}$$

$$\dot{x}_2 = -\left(\frac{c}{m}\right)x_2 - \left(\frac{k}{m}\right)x_1 + \left(\frac{1}{m}\right)p \tag{2.19}$$

Now, we can put Eqs. (2.19) and (2.16) into matrix form as:

$$\underbrace{\begin{bmatrix} \dot{x}_1 \\ \dot{x}_2 \end{bmatrix}}_{\dot{x}} = \underbrace{\begin{bmatrix} 0 & 1 \\ -\dfrac{k}{m} & -\dfrac{c}{m} \end{bmatrix}}_{A} \underbrace{\begin{bmatrix} x_1 \\ x_2 \end{bmatrix}}_{x} + \underbrace{\begin{bmatrix} 0 \\ \dfrac{1}{m} \end{bmatrix}}_{B} u \tag{2.20}$$

$$y = \underbrace{[1 \ 0]}_{C} \underbrace{\begin{bmatrix} x_1 \\ x_2 \end{bmatrix}}_{x} + \underbrace{[0]}_{D} u \tag{2.21}$$

2.2 Linear Systems

2.2.1 The Fundamental Theorem for Linear Systems

Theorem 2.1 *The Fundamental Theorem for Linear Systems*
Let A be a $n \times n$ matrix (Eq. (2.22)). Then for a given $x_0 \in \mathcal{R}^n$, the initial value problem (IVP)

$$\dot{x} = Ax$$
$$x(0) = x_0 \tag{2.22}$$

has a unique solution given by

$$x(t) = e^{tA}x_0 \tag{2.23}$$

Proof If $x(t) = e^{tA}x_0$, then

$$\dot{x} = Ae^{tA}x_0 = Ax \tag{2.24}$$

Moreover

$$x(0) = x_0 \tag{2.25}$$

Thus $x(t) = e^{tA}x_0$ is a solution of the IVP. Let $y(t)$ be any solution of the IVP and set

$$w(t) = e^{-At}y(t) \tag{2.26}$$

Then

$$\dot{w} = -e^{-At}Ay(t) + e^{-At}\dot{y}(t)$$
$$= -e^{-At}Ay(t) + e^{-At}y(t)$$
$$= 0$$

Thus, $w(t)$ is a constant. But $x(0) = y(0) = x_0$. Hence $w(t) = x_0$ and $y(t) = e^{tA}x_0$.

2.2.2 Linear Systems in R^2

Consider the linear system given in Eq. (2.22)

$$\dot{x} = Ax \qquad (2.27)$$

$x \in \mathfrak{R}^2$, A is a real matrix of 2×2

Because it is possible to calculate the exponential matrix of 2×2, the solution is obtained explicitly for any system given by (2.27) and qualitative information is extracted.

Consider the linear system given by (2.27), with det $A \neq 0$. It is possible to determine when a linear system has a saddle-type phase portrait, node, focus or center at the origin. The above is obtained from the analysis of the eigenvalues of matrix A, which are found by solving the characteristic equation:

$$\lambda^2 - (trace(A))\lambda + \det A = 0 \qquad (2.28)$$

Theorem 2.2 If $\tau = \text{trace}(A)$ and $\delta = \det A$. Consider the system linear (2.23)

  i. If $\delta < 0 \rightarrow$ then (2.27) has a saddle point at the origin.
  ii. If $\delta > 0$ and $\tau = 0 \rightarrow$ then (2.27) has a center at the origin.
  iii. If $\delta > 0$, $\tau \neq 0$, and $\tau^2 - 4\delta < 0 \rightarrow$ then (2.27) has a node at the origin; and is stable if $\tau < 0$ or unstable if $\tau > 0$.
  iv. If $\delta > 0$ and $\tau^2 - 4\delta \geq 0 \rightarrow$ then (2.27) has a focus on the origin; and is stable if $\tau < 0$ or unstable if $\tau > 0$.

2.2.3 Complex Eigenvalues

In Eq. (2.27), we obtained the solutions to a homogeneous linear system with constant coefficients to solve a system of n equations (written in matrix form as (2.27)) we must find n linearly independent solutions $x_1, ..., x_n$.

Case 1

Where A has n real and distinct eigenvalues, we have already solved the system by using the solutions $e^{\lambda_i t} v_i$, where λ_i and v_i are the eigenvalues and eigenvectors of A.

Case 2

Where A has complex eigenvalues. We know that under the assumption that the roots of its characteristic equation (2.27)

$$|A - \lambda I| = 0 \qquad (2.29)$$

Theorem 2.3 The scalar λ is an eigenvalue of the $n \times n$ matrix A if and only if $\det(A-\lambda I) = 0$.

if there are complex eigenvalues. Since the characteristic equation has real coefficients, its complex roots must occur in conjugate pairs:

$$\lambda = a + bi \; \lambda = a - bi$$

Since λ is complex, the a will also be complex, and therefore the eigenvector v corresponding to λ will have complex components. Putting together the eigenvalue and eigenvector gives us formally the complex solution

$$x = e^{(a-bi)t}v \tag{2.30}$$

Here, the following theorem tells us to just take the real and imaginary parts of (2.30) (this theorem is exactly analogous to what we did with ODEs).

Theorem 2.4   Given a system (2.27), where A is a real matrix. If $x = x_1 + ix_n$ is a complex solution, then its real and imaginary parts x_1, x_2 are also solutions to the system [9].

Now let us write the eigenvector split into real and imaginary parts, as

$$v = a + bi \tag{2.31}$$

If we also write our eigenvalues with real and imaginary parts as $r = \lambda + \varphi i$ then one solution can be rewritten as follows:

$$(a + bi)e^{(\lambda+\varphi i)t} = (a + bi)e^{\lambda t}(\cos(\varphi t) + i\sin(\varphi t)) \tag{2.32}$$

2.2.4   Multiple Eigenvalues

It may be that a matrix A has some "repeated" eigenvalues. That is, the characteristic equation $\det(A - \lambda I) = 0$ may have repeated roots. However, in one case this is important, namely, when the matrix A has multiple eigenvalues. In that case, we will get only one vector from the multidimensional invariant subspace of these multiple eigenvalues.

Theorem 2.5   Suppose the $n \times n$ matrix P has n real eigenvalues (not necessarily distinct), $\lambda_1, \lambda_2, \ldots, \lambda_n$, and there are n linearly independent corresponding eigenvectors $\vec{v}_1, \vec{v}_2, \ldots, \vec{v}_n$. Then the general solution $\dot{\vec{x}} = P\vec{x}$ can be written as

$$\vec{x} = c_1\vec{v}_1 e^{\lambda_1 t}, c_2\vec{v}_2 e^{\lambda_2 t} + \ldots + c_n\vec{v}_n e^{\lambda_n t}. \tag{2.33}$$

In other words, the hypothesis of the theorem could be stated as saying that if all the eigenvalues of P are complete, then there are n linearly independent eigenvectors and thus we have the given general solution [10].

2.3   Nonlinear Dynamical Systems and its Analysis

2.3.1   Preliminary Concepts and Definitions

Nonlinear dynamical system (NDS) theory which includes specific techniques and concepts such as chaos, dissipative structures, bifurcation, catastrophe theory, and fractals shows that extreme complexity can also arise due to the nonlinear dynamics and couplings of relatively simple systems represented by relatively small equation systems (deterministic complexity). If the system (plant or bioprocesses) evolves with time, in such a way that the states of the system at time t depend upon the states of the system at earlier times, they are called dynamical systems. In other words, a dynamical system consists of a set of possible states,

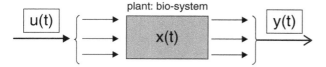

Figure 2.4 Dynamical system.

together with a rule that determines the present states in terms of past states. According to the character of the state variables dynamical systems this can be classified as a discrete dynamical system or as a continuous dynamical system [1].

We consider a dynamical system (Figure 2.4).

Where:

i. y is the output: represents what is "measurable" from outside the system.
ii. x is the state of the system: characterizes the state of the system.
iii. u is the control input: makes the system move.

2.3.1.1 Continuous Dynamical Systems

A continuous dynamical system is represented by a differential equation

$$\dot{x} = \frac{dx}{dt} = f(x, t); x \in R^n \forall t \in R \tag{2.34}$$

possessing a unique solution $x(t, t_0) = x(t)$ satisfying the condition $x(t_0) = x_0$ in bioprocesses $x \in \Re^n$ is the vector of the state variables for example:

$$x = [S_{total}, S_{sus}, S_{res}, C, E, G, B]^T \in \Re^7_+ \text{ concentrations in the bioprocesses.}$$

2.3.1.2 Phase Space and Phase Portrait

A set of phase variables of a system is a minimal set of variables which fully describes the state of the bio-system (systems of ordinary differential equations, EDOs). Phase space is the space generated by the phase variables (2D) i.e. phase space is the space generated by the generalized coordinates and generalized momenta of physical, chemical, and biological systems. The state of a system at any time is represented by a point in the system's phase space. Change of a system state over time is represented by a trajectory in the phase space. In other words, a trajectory is the path of an object in phase space as a function of time. A phase portrait is the collection of all possible trajectories of the system. Also, the phase portrait for a differential equation is the collective graph of trajectories of the differential equation in the phase plane [11].

2.3.1.3 Trajectories of Autonomous and Non-Autonomous Systems

If two trajectories intersect, it indicates that at the point of intersection, the system can evolve in more than one possible direction (autonomous systems cannot intersect each other). This violates the existence of a unique solution of an autonomous dynamical system. On the other hand, non-autonomous system trajectories can have self-intersection and two different trajectories can intersect at a later time.

2.3.1.4 The Vector Field

A vector field arises in a situation where, for some reason, there is a direction and a magnitude assigned to each point of the space or of a surface (dynamical system). A vector field is a smooth function $f : \Omega \subset \mathfrak{R}^n \to \mathfrak{R}^n$. In this frame, the vector field is steady: it does not depend on time.

For example: any problem implying the gradient Δh of a scalar function defines a vector field; if h is a smooth function, then the gradient

$$\Delta h = \left(\frac{\partial h}{\partial x_1}, \frac{\partial h}{\partial x_2}, \cdots, \frac{\partial h}{\partial x_n} \right) \tag{2.35}$$

One of the applications of vector fields is the visualization of the solution of ODEs. Differential equations define a vector field at every point in the phase space. The solution of a differential equation with prescribed initial condition follows the flow of vectors [11–13].

2.3.1.5 Lipschitz Condition

A function $f(x)$ is said to be locally Lipschitz on a domain (open and connected set) $D \subset \mathfrak{R}^2$ if each point of D has a neighborhood D such that f satisfies

$$\|f(x) - f(y)\| \leq \ell \|x - y\|, \quad \forall x.y \in D_0 \tag{2.36}$$

with the same Lipschitz constant ℓ. The same terminology is extended to a function $f(x, t)$, provided that the Lipschitz constant holds uniformly in t for all t in a given interval.

Remark For $f : \mathfrak{R} \to \mathfrak{R}$, we have

$$\frac{|f(x) - f(y)|}{x - y} \leq \ell \tag{2.37}$$

which means that a straight line joining any two points of f cannot have a slope whose absolute value is greater than ℓ. Lipchitz continuity (local and global): Understanding Lipschitz continuity is necessary to realize the existence and uniqueness theory of ODE. A function $f(x, y)$ is said to be locally Lipschitz or locally Lipschitz continuous at a point $y_0 \in D$ (open and connected set) if y_0 has a neighborhood D_0 such that $f(x, y)$ satisfies

$$|f(x, y_1) \text{-} f(x, y_2)| < \ell |y_1 - y_2| \forall (x, y_1)(x, y_2) \in D_0 \tag{2.38}$$

Where ℓ is fixed over the neighborhood D_0.

A function is said to be locally Lipschitz in a domain D if it is locally Lipschitz at every point of the domain D. We denote the set of all locally Lipschitz functions by ℓ, and symbolically we say $f \in \ell$, if f is Lipschitz locally [11–13]. Moreover, a function is said to be globally Lipschitz (or $f \in$ set of globally Lipschitz functions) if

$$|f(x, y_1) - f(x, y_2)| < \ell |y_1 - y_2| \ \forall (x, y_1), (x, y_2) \in D \tag{2.39}$$

with the same Lipschitz constants ℓ for entire D.

2.3.2 Existence-Uniqueness Theorem

From the above discussions, it is quite clear that first-order linear ODEs are easy to solve because there are straightforward methods for finding solutions of such equations (with continuous coefficients). In contrast, there is no general method for solving even first-order nonlinear ODEs. Some typical nonlinear first-order ODEs can be solved by Bernoulli's method (1697), a method of separation of variables, of variation of parameters, and using integrating factors [11].

The following issues are remarkable. Arguments based on continuity of functions are common in dynamical system analysis, as an alternative, continuity arguments can be used to show that certain qualitative conditions cannot be satisfied for a class of systems.

In general formulae for solving nonlinear ODEs have two important consequences:

i. Firstly, methods which yield approximate solution (numerical) and give qualitative information about solutions assume greater significance for nonlinear equations.
ii. Secondly, questions dealing with the existence and uniqueness of solutions become important [11].

Regarding the existence and uniqueness of solutions of first order ODE theory the following questions naturally arise:

$$\dot{y} = f(x, y), \quad y(x_0) = y_0 \tag{2.40}$$

i. Under what conditions can we be sure that a solution to (2.34) exists?
ii. Under what conditions can we be sure that there is a unique solution to (2.34)?
iii. For which values of x does, the solution to (2.34) exist (the interval of existence)?

Theorem 2.6 *Picard–Lindelof*
If f and $\frac{\partial f}{\partial y}$ are continuous functions of x, y in some rectangle $:\{(x, y): \alpha < x < \beta;\, \gamma < y < \delta\}$ containing the point (x_0, y_0) then in some interval $x_0 - \delta < x < x_0 + \delta (\delta > 0)$ there exists a unique solution of IVP (2.34) (Figure 2.5).

Let us state the existence and uniqueness theorem which Lipschitz knew. Let $f(x, y)$ in (2.34) be continuous on a rectangle $\Theta = \{(x, y): x_0 - \delta < x < x_0 + \delta;\, y_0 - \rho < y < y_0 + \rho\}$ then there exists a solution in Θ, furthermore, if $f(x, y)$ is Lipschitz continuous with respect

Figure 2.5 Showing initial condition (*x0,y0*) enclosed in neighborhoods R and D.

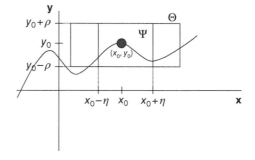

to y on a rectangle Ψ (possibly smaller than Θ) given as $\Psi = \{(x, y) : x_0 - \eta < x < x_0 + \eta;\ y_0 - \rho < y_0 + \rho, \eta < \delta\}$ the solution in R shall be unique (Figure 2.5).

Theorem 2.7   Local Existence and Uniqueness
Let $f(x, t)$ be piecewise continuous in t and satisfy the Lipschitz condition

$$f(x, t)\text{-}f(y, t) \leq \ell \|x - y\| \tag{2.41}$$

$$\forall x, y \in B = \{x \in \mathfrak{R}^n : \|x - x_0\| \leq r\},\ \forall t \in [t_0, t_1]$$

Then, there exists some $\delta > 0$ such that the state equation $\dot{x} = f(x, t)$ with $x(t_0) = x_0$ has a unique solution over $[t_0, t_0 + \delta]$.

Theorem 2.8   Global Existence and Uniqueness
Suppose that $f(t, x)$ is piecewise continuous in t and satisfies

$$\|f(x, t)\text{-}f(y, t)\| \leq \ell \|x - y\|, \forall x, y \in \mathfrak{R}^n, \forall t \in [t_0, t_1] \tag{2.42}$$

Then, the state equation $\dot{x} = f(x, t)$, with $x(t_0) = x_0$, has a unique solution over

$$[t_0, t_1].$$

Lemma 2.1   If $f(x, t)$ and $\left[\frac{\partial f}{\partial x}\right]_{(x,t)}$ are continuous on $[a, b] \times \Omega$ for some domain $\Omega \subset \mathfrak{R}^n$, then f is locally Lipschitz in x on $[a, b] \times \Omega$.

Lemma 2.2   If $f(x, t)$ and $\left[\frac{\partial f}{\partial x}\right]_{(t,x)}$ are continuous on $[a, b] \times \mathfrak{R}^n$, then f is globally Lipschitz in x on n $[a, b] \times \mathfrak{R}^n$ if and only if $\left[\frac{\partial f}{\partial x}\right]$ is uniformly bounded on $[a, b] \times \mathfrak{R}^n$.

Theorem 2.9   Let $f(x, t)$ be piecewise continuous in t and locally Lipschitz in x for all $t \geq t_0$ and all $x \in \Omega \subset \mathfrak{R}^n$. Let W be a compact subset of Ω, $x_0 \in W$ and suppose it is known that every solution of

$$\dot{x} = f(x, t), x(t_0) = x_0 \tag{2.43}$$

lies entirely in W. Then, there is a unique solution that is defined for all $t \geq t_0$.

2.3.2.1   Algebraic Properties of Lipschitz Continuous Functions
We have so far realized the importance of Lipschitz continuity in the existence and uniqueness theory of IVP of ODE. We now discuss some algebraic properties of ℓ, functions. Suppose $f_1, f_2 \in I$ with Lipschitz constant ℓ_1, ℓ_2 on the interval I. Then for two points y_1, $y_2 \in I$

$$\begin{aligned}
&|(f_1 + f_2)(y_1) - (f_1 + f_2)(y_2)| \\
&= |f_1(y_1) + f_2(y_1) - f_1(y_2) - f_2(y_2)| \\
&\leq |f_1(y_1) + f_1(y_2)| + |f_2(y_1) - f_2(y_2)| \\
&\leq \ell_1 |y_1 - y_2| + \ell_2 |y_1 - y_2| \\
&= (\ell_1 + \ell_2) |y_1 - y_2|
\end{aligned} \tag{2.44}$$

Remark

i. Recall that the existence and uniqueness theorem for ODEs guarantees that solutions are defined only for t close to 0. Therefore, our flow $\omega(X, t)$ is defined only for t in some neighborhood $G \subset B$ of zero. This neighborhood, in general, depends on X, so we can write $G = G(X)$. The flow should, therefore, be called the local flow, to indicate that solutions are only defined locally (in t). It would also be more correct to say that φ_t is defined on the set

$$\Omega = \{(X, t) \in \mathfrak{R} \times \mathfrak{R}^n : t \in G(X)\} \tag{2.45}$$

not on all of $\mathfrak{R} \times \mathfrak{R}^n$.

ii. Given a (smooth) collection of maps $\omega_t : \mathfrak{R}^n \to \mathfrak{R}^n$ satisfying $\omega_0 = 0$ identity and $\omega_{s+t} = \omega_s \circ \omega_t$, we can always recover the vector field F so that ω_t is the flow of F. Just differentiate with respect to t:

$$F(X) = \left.\frac{d}{dx}\right|_{t=0} \varphi_t(X) \tag{2.46}$$

2.3.3 Dependence on Initial Conditions and Parameters

Consider the following nominal model

$$\dot{x} = f(x, y, \pi_0) \tag{2.47}$$

where $\pi_0 \in \mathfrak{R}^p$ denotes the nominal vector of constant parameters of the model and $x \in \mathfrak{R}^n$ is the state.

- Let $y(t)$ be a solution of (2.47) that starts at $y(t_0) = y_0$ and is defined on the interval $[t_0, t_1]$.
- Let $z(t)$ be a solution of $\dot{x} = f(x, y, \pi_0)$ defined on $[t_0, t_1]$ with $z(t_0) = z_0$.

Or is, in other words, the solution continuous dependent on the initial condition and parameter π

$$\forall_{\delta > 0} \exists_{\delta > 0} : z_0 - y_0 < \delta, \|\pi - \pi_0\| < \delta \Rightarrow z(t) - y(t) < \varepsilon, \forall_{t \in [t_0, t_1]}$$

That is, if we consider the system of differential equations that depend on a parameter

$$\dot{x} = f(x, t, \pi) \tag{2.48}$$

here $f : B_\pi \to \mathfrak{R}^n$, $B_v \subset \mathfrak{R}^{n+m+1}$ is a region. Let us suppose that the region G_π satisfies the condition of Lipschitz locally:

$$f \in \ell(B_\pi)_{loc}$$

By the existence and uniqueness theorem $B = \{(x, t) : (x, t, \pi) \in B_\pi\}$ is the region of uniqueness for the system (2.48) for each π. Let us consider the function $x(x_0, t_0, t, \pi)$, system solution (2.48) with the initial conditions x_0, t_0 defined in the set

$$B_\pi = \{(x_0, t_0, t, \pi) : (x_0, t_0, \pi) \in B_\pi, t \in I\left(x_0, t_0, \pi\right)\}, \tag{2.49}$$

Here $I(x_0, t_0, \pi)$ it is the maximum interval of existence.

Theorem 2.10 Under the assumptions made, B_π is a region and $x(x_0, t_0, \pi)$ is continuous in D_π.

Differentiability of the solutions with respect to the initial data and parameters.

Now, suppose that $\frac{\partial f}{\partial x}$ y $\frac{\partial f}{\partial \pi}$ exist and are continuous in the region B_π. Consider the differentiability of the solution $x(x_0, t_0, t, \pi)$ in the region B_π.

Theorem 2.11 *Continuity of Solutions*

Let $f(x, t, \pi)$ be continuous in (x, t, π) and locally Lipschitz in x on $[t_0, t_1] \times D \times \{\|\pi - \pi_0\| \leq c\}$ where $D \subset \mathfrak{R}^n$ is an open connected set. Let $f(t, \pi_0)$ be a solution of

$$\dot{x} = f(x, t, \pi_0); \tag{2.50}$$

with $y(t_0, \pi_0) = y_0 \in D$. Suppose $y(t, \pi_0)$ is defined and belongs to D for all $t \in [t_0, t_1]$. Then, given $\varepsilon > 0$, there is $\delta > 0$ such that if

$$\|z_0 - y_0\| < \delta, \|\pi - \pi_0\| < \delta$$

then there is a unique solution $z(t, \pi)$ of $\dot{x} = f(t, x, \pi)$ defined on $[t_0, t_1]$, with $z(t_0, \pi) = z_0$ such that $\|z(t, \pi - y(t, \pi_0))\| < \varepsilon, \forall t \in [t_0, t_1]$.

2.3.4 The Flow Defined by a Differential Equation

The definition of the flow, φ_t, of the nonlinear system is

$$\dot{x} = f(x) \tag{2.51}$$

In the following definition, we denote the maximal interval of the existence (α, β) of the solution $\varphi(x_0, t)$ of the IVP

$$\dot{x} = f(x) x(0) = x_0 \tag{2.52}$$

by $I(x_0)$ since the endpoints α and β of the maximal interval are generally dependent on x_0.

Definition 2.1 *Flow*

Let Θ be an open subset of $\mathfrak{R}^n$ and assume that $f \in C^1(E)$. For $x_0 \in \Theta$, (2.52) defined its maximal interval of existence $I(x_0)$. Then for $t \in I(x_0)$, the set of mappings φ_t defined by

$$\varphi_t(x_0) = \varphi(x_0, t)$$

is called the flow of the differential Eq. (2.51) or the flow defined by the differential Eq. (2.51); φ_t, also referred to as the flow of the vector field $f(x)$.

We define the set $\Omega \subseteq \mathfrak{R} \times E$ as

$$\Omega = \{(x_0, t) \in \mathfrak{R} \times \Theta : t \in I(x_0)\}.$$

2.3.4.1 Differential Flow

Consider a time-invariant autonomous ODE

$$\dot{x}(t) = f(x, t) \tag{2.53}$$

where $f : \mathfrak{R}^{n+1} \mapsto \mathfrak{R}^n$ satisfies the Lipschitz constraint

$$|f(x_1) - f(x_2)| \leq \ell |x_1 - x_2| \tag{2.54}$$

on every bounded subset of $\mathfrak{R}^n$.

If $x : \Omega \to \mathfrak{R}$ is a continuous function defined on an open subset $\Omega \subset \mathfrak{R} \times \mathfrak{R}^n$, with π considered a parameter, $t \to x(t, \pi)$ defines a family of smooth curves in $\mathfrak{R}^n$.

When t is fixed, $\pi \to x(t, \pi)$ defines a continuous map from an open subset of $\mathfrak{R}^n$ and with values in $\mathfrak{R}^n$.

Note that $x(t_1, x(t_2, \pi)) = x(t_1 + t_2, \pi)$ whenever $x(t_2, x)/\infty$.

The function $x : \Omega \mapsto \mathfrak{R}^n$ is sometimes called the "differential flow" defined by (2.53).

2.3.5 Equilibrium Points

A point $x_0 \in \mathfrak{R}^n$ is called an equilibrium of (2.53) when $f(x_0) = 0$, i.e. $x = (x_0, t) \equiv x_0$ is a constant solution of (2.53).

Definition 2.2 An equilibrium x_0 of (2.53) is called asymptotically stable if the following two conditions are satisfied:

i. there exists $d > o$ such that $x(t, \overline{x}) \to \overline{x}_0$ as $t \to \infty$ for all $\overline{x}$ satisfying $|\overline{x}_0 - \overline{x}| < d$;
ii. for every $\varepsilon > 0$ there exists $\delta > 0$ such that $|x(t, x) - x_0| < \varepsilon$ whenever $t \geq 0$ and $|\overline{x} - \overline{x}_0| < \delta$.

2.3.5.1 Equilibrium
Consider and autonomous dynamical system (2.53)

$$\frac{dx}{dt} = \dot{x} = f(x, t) \tag{2.55}$$

The state x_{eq} is called equilibrium when it satisfies the following equation

$$0 = f(x_{eq}, t) \tag{2.56}$$

For non-autonomous systems:

$$\dot{x} = f(t, x, u), \tag{2.57}$$

a fixed input u^{eq} is called the equilibrium with respect to input u^{eq} when it satisfies the following equation,

$$0 = f(x_{eq}, t, u_{eq}) \tag{2.58}$$

Energy systems do not always present themselves directly as mathematical models in state space form with their equilibria explicitly defined. More often, we arrive at a collection of ODEs and algebraic equations from first principles. However, we can usually carefully select the state variables to form a state-space representation and compute equilibria, as shown by the following example.

An equilibrium point $\dot{x}$ of a continuous dynamical system is the point in the phase space where phase space velocity is zero.

$$\text{i.e. } \frac{dx}{dt}\bigg|_{x=x_{eq}} = 0 \tag{2.59}$$

$$\text{i.e. } f(x_{eq}) = 0 \tag{2.60}$$

Notice that the equilibrium points of a dynamical system are solutions of the ODE.

Example:

$$\frac{dx}{dt} = x(1 - x) \text{ has equilibrium points 0 and 1.}$$

Definition 2.3

i. An equilibrium point x_0 of (2.55) is stable if, for all $\varepsilon > 0$, there exists $\delta > 0$ such that for all $x \in D_\delta(x_0)$ and $t \geq 0$, we have $\phi_t(x) \in N_c(x_0)$

ii. An equilibrium point x_0 of (2.55) is unstable (i.e. not stable).

iii. An equilibrium point x_0 of (2.55) is asymptotically stable if it is stable and if there exists a $\delta > 0$ such that for all $x \in D_\delta(x_0)$ we have $lim_{t \to \infty} \phi_t(x) = x_0$.

Remarks

i. The above limit being satisfied does not imply that x_0 is stable (why?).

ii. From H–G theorem and stable manifold theorem, it follows that hyperbolic equilibrium points are either asymptotically stable (sinks) or unstable (sources or saddles).

iii. If x_0 is stable then no eigenvalue of $Df(x_0)$ has a positive real part; why?

iv. x_0 is stable but not asymptotically stable, then x_0 is a non-hyperbolic equilibrium point [14, 15]

2.3.6   The Hartman–Grobman Theorem

Definition 2.4

i. Let Λ be a metric space (such as $\mathfrak{R}^n$) and let A and B be subsets of X. A homeomorphism of A onto B is a continuous one-to-one map of A onto B, $h : A \to B$, such that $h^{-1} : B \to A$ is continuous.

ii. The sets A and B are called homeomorphic or topologically equivalent if there is a homeomorphism of A onto B.

Theorem 2.12   *The Hartman–Grobman Theorem*
Let $f \in C^1(\Lambda)$ where Λ is an open subset of R^n containing the origin, and φ_t the flow of (2.55). Suppose that $f(0) = 0$ and that $Df(0)$ has no eigenvalues with a zero real part [16, 17]. Then there is a homeomorphism H of an open set U containing the origin onto a set V containing the origin such that for each $x_0 \in U$, there is an open interval $I_0 \subset \mathfrak{R}$ containing zero such that for all $x_0 \in U$ and $t \in I_0$

$$H \circ \varphi_t(x_0) = e^{At} H(x_0) \tag{2.61}$$

Theorem 2.13   If an nth-order system of differential equations has an equilibrium $\vec{c}$ with linearization matrix A, and if A has no zero or pure imaginary eigenvalues, then the phase portrait for the system near the equilibrium is obtained from the phase portrait of the linearized system $D\vec{x} = A\vec{x}$ via a continuous change of coordinates [16, 17].

Theorem 2.14 For a second-order system with an equilibrium that satisfies the hypotheses of the Hartman–Grobman Theorem, the continuous change of coordinates can, in fact, be taken to be continuously differentiable and such that the derivative at the equilibrium is the identity.

2.3.7 The Stable Manifold Theorem

$$\dot{x} = f(x) \tag{2.62}$$

$$\dot{x} = Df(x_0)x \tag{2.63}$$

We assume that the equilibrium point x_0 is located at the origin.

Theorem 2.15 *The Stable Manifold Theorem*

Let Λ be an open subset of $\mathfrak{R}^n$ containing the origin, let $f \in C^{l1}(\Lambda)$, and let φ_t be the flow of the nonlinear system (2.62). Suppose that $f(0) = 0$ and that $Df(0)$ has k eigenvalues with negative real part and $n - k$ eigenvalues with positive real part. Then there exists a $k-$ dimensional manifold S tangent to the stable subspace Λ^s of the linear system (2.63) at 0 such that for all
$t \leq 0$, $\varphi_t(S) \subset S$ and for all $x_0 \in S$,

$$\lim_{t \to \infty} \varphi_t(x_0) = 0 \tag{2.64}$$

and there exists an $n - k$ differentiable manifold U tangent to the unstable subspace Λ^u of (2.63) at 0 such that for all $t \leq 0$, $\varphi_t(U) \subset U$ and for all $x_0 \in U$,

$$\lim_{t \to -\infty} \varphi_t(x_0) = 0 \tag{2.65}$$

Note:

As in the examples, since $f \in C^{l1}(\Lambda)$ and $f(0) = 0$, then system (2.62) can be written as

$$\dot{x} = Ax + F(x) \tag{2.66}$$

where $A = Df(0)$, $F(x) = f(x) - Ax$, $F(0) = 0$ and $DF(0) = 0$.

Furthermore, we want to separate the stable and unstable parts of the matrix, i.e. choose a matrix C such that

$$B = C^{-1}AC = \begin{bmatrix} P & 0 \\ 0 & Q \end{bmatrix} \tag{2.67}$$

where the eigenvalues of the $k \times k$ matrix P have a negative real part, and the eigenvalues of the $(n - k) \times (n - k)$ matrix Q have a positive real part. The transformed system $(y = C^{-1}x)$ has the form

$$\dot{y} = By + C^{-1}F(Cy)$$
$$\dot{y} = By + G(y) \tag{2.68}$$

Theorem 2.16 *The Center Manifold*

Let $f \in C^r(E)$ where E is an open subset of $\mathfrak{R}^n$ containing the origin and $r \geq 1$. Suppose that $f(0) = 0$ and that $Df(0)$ has k eigenvalues with a negative real part, j eigenvalues with a positive real part, and $m = n - k - j$ eigenvalues with zero real part. This theorem can be used to determine the near non-hyperbolic equilibrium point.

The center manifold theorem allows us to determine the local behavior of the system by looking at a lower dimensional manifold, i.e. instead of working with an n-dimensional system, we can just deal with a c-dimensional one (i.e. we reduce the full n-dimensional system into a c-dimensional system) [18].

2.3.8   Saddles, Nodes, Foci, and Centers

Consider that the nonlinear system in (2.62) was said to have a saddle, a sink or a source at a hyperbolic equilibrium point x_0 if the linear part of f at x_0 had eigenvalues with both positive and negative real parts, only had eigenvalues with negative real parts, or only had eigenvalues with positive real parts, respectively.

Similarly, a source of (2.62) is either an unstable node or focus of (2.62) as defined below. Finally, we define centers and center-foci for the nonlinear system (2.62) and show that, under the addition of nonlinear terms, a center of the linear system (2.22) may become a center, a center-focus, or a stable or unstable focus of (2.62).

Before defining these various types of equilibrium points for planar systems (2.62), it is convenient to introduce polar coordinates (r, θ) and to rewrite the system (2.62) in polar coordinates.

The nonlinear system (2.62) can then be written as

$$\dot{x} = P(x, y);$$
$$\dot{y} = Q(x, y); \qquad\qquad (2.69)$$

If we let $r^2 = x^2 + y^2$ and $\theta = \tan^{-1}(y/x)$, then we have

$$r\dot{r} = x\dot{x} + y\dot{y}$$

and

$$r^2\dot{\theta} = x\dot{y} - y\dot{x} :$$

It follows that for $r > 0$; the nonlinear system (2.69) can be written in terms of polar coordinates as

$$\dot{r} = P(r\cos\theta, r\sin\theta)\cos\theta + Q(r\cos\theta, r\sin\theta)\cos\theta$$
$$r\dot{\theta} = Q(r\cos\theta, r\sin\theta)\cos\theta - P(r\cos\theta, r\sin\theta)\cos\theta \qquad (2.70)$$

$$\frac{dr}{d\theta} = \frac{[P(r\cos\theta, r\sin\theta)\cos\theta + Q(r\cos\theta, r\sin\theta)\sin\theta]}{Q(r\cos\theta, r\sin\theta)\cos\theta - P(r\cos\theta, r\sin\theta)\sin\theta} \qquad (2.71)$$

Writing the system of differential Eq. (2.69) in polar coordinates will often reveal the nature of the equilibrium point or critical point at the origin. We assume that $x_0 \in \mathcal{R}^2$ is an isolated equilibrium point of the nonlinear system (2.69) which has been translated to the origin; $r(t, r_0, \theta_0)$ and $\theta(t, r_0, \theta_0)$ will denote the solution of the nonlinear system (2.70) with $r(0) = r_0$ and $\theta(0) = \theta_0$ [19].

Definition 2.5   The origin is called a center for the nonlinear system (2.62) if there exists $\delta > 0$ such that every solution curve of (2.62) in the deleted neighborhood $N_\delta(0) - \{0\}$ is a closed curve with 0 in its interior.

Definition 2.6   The origin is called a center-focus for (2.62) if there exists a sequence of closed solution curves T_n, with T_{n+1} in the interior of T_n such that $T_n \to 0$ as $n \to \infty$ and such that every trajectory between T_n and T_{n+1} spirals toward T_n or T_{n+1} as $t \to \pm\infty$.

Definition 2.7   The origin is called a stable focus for (2.62) if there exists $\delta > 0$ such that for $0 < r_0 < \delta$ and $\theta \in \mathfrak{R}$, $r(t, r_0, \theta_0) \to 0$ and $|\theta(t, r_0, \theta_0)| \to \infty$ as $t \to \infty$. It is called an unstable focus if $r(t, r_0, \theta_0) \to 0$ and $|\theta(t, r_0, \theta_0)| \to \infty$ as $t \to \infty$. Any trajectory of (2.62) which satisfies $r(t) \to 0$ and $|0(t)| \to \infty$ as $t \pm \infty$ is said to spiral toward the origin as $t \pm \infty$.

Definition 2.8   The origin is called a stable node for (2.62) if there exists a $\delta > 0$ such that for $0 < r_0 < \delta$ and $\theta \in \mathfrak{R}$, $r(t, r_0, \theta_0) \to \infty$ as $t \to \infty$ and $\lim_{t \to \infty} \theta(t, r_0, \theta_0)$ exists; i.e. each trajectory in a deleted neighborhood of the origin approaches the origin along a well-defined tangent line as $t \to \infty$.

The origin is called an unstable node if $r(t, r_0, \theta_0) \to 0$ and $\lim_{t \to -\infty} \theta(t, r_0, \theta_0)$ exists for all $r_0 \in (0, \delta)$ and $\theta \in \mathfrak{R}$

The origin is called a proper node for (2.62) if it is a node and if every ray through the origin is tangent to some trajectory of (2.62).

Theorem 2.17   Let E be an open subset of $\mathfrak{R}^2$ containing the origin and let $f \in C^1(E)$. If the origin is a hyperbolic equilibrium point of the nonlinear system $r(t, r_0, \theta_0) \to 0$ then the origin is a (topological) saddle for this nonlinear system if and only if the origin is a saddle for the linear system

$$\dot{x} = Ax; \tag{2.72}$$

with $A = Df(0)$.

Theorem 2.18   Let E be an open subset of $\mathfrak{R}^2$ containing the origin and let $f \in C^1(E)$ with $f(0) = 0$.

If the nonlinear system

$$\dot{x} = f(x) \tag{2.73}$$

is symmetric with respect to the $x-$ axis or the $y-$ axis, and if the origin is a center for the linear system

$$\dot{x} = Ax \tag{2.74}$$

with $A = Df(0)$, then the origin is a center for the nonlinear system

$$\dot{x} = f(x) \tag{2.75}$$

2.3.9 Center Manifold Theory

i. Center manifold theory allows us to reduce the dimension of a problem, you will most likely still be left with a nonlinear system.

ii. Normal form theory can be used to "simplify" the nonlinear system by removing as much nonlinearity as possible. This involves nonlinear coordinate transformation.

iii. Local bifurcation theory uses the above techniques to determine when the system changes qualitatively as parameters are varied.

2.4 Stability Theory via Lyapunov Approach

2.4.1 Stability Notions

In this section, we will investigate the concept of Lyapunov stability (defined in 1892). We will define the following notions: stable, asymptotically stable, equilibrium point. We show that the eigenvalues of the matrix of some linear system with constant coefficients decide on the stability of the system. The stability of the equilibrium points of nonlinear systems is studied in two ways: via linearization and via Lyapunov functions. Lyapunov stability: an intuitive approach to analyzing the stability and convergence of dynamic systems (SL-SNL) without explicitly numerical solutions of their differential equations (phenomenological bio-process). This method forms the basis of much of modern nonlinear control theory and also provides a theoretical justification for using local linear control techniques [20–23].

2.4.1.1 Stability

We consider the differential system

$$\dot{x} = f(x, t) \tag{2.76a}$$

where

$$f \in C^1(\mathfrak{R}^n \to \mathfrak{R}^n)$$

- $x_{eq} \in \mathfrak{R}^n$ is such that $f(x_{eq}) = 0$
- $f(x, t)$ is Lipschitz continuous with respect to x
- $f(x, t)$ uniformly in t
- $f(x, t)$ piecewise continuous in t
- (2.76a) we use state equation of the form

$$\dot{x} = \underbrace{f(x, \mu, t)} \tag{2.76b}$$

x : *state variable*
μ : *control input*

- The system (2.76b) is autonomous if f does not depend explicitly on time t. Otherwise, the system (2.76b) is non-autonomous.

Definition 2.9   A point x_{eq} is an *equilibrium point* of (1) if $f(x_{eq}, t) \equiv 0$, we say an equilibrium point is *locally stable* if all solutions which start near x_{eq} (meaning that the initial conditions are in a neighborhood of x_{eq} remain near x_{eq} for all time). The equilibrium point x_{eq} is said to be *locally asymptotically stable* if x_{eq} is locally stable and, additionally, all solutions starting near x_{eq} tend toward x_{eq} as $t \to \infty$ [24].

Definition 2.10   The equilibrium x_{eq} of (2.76a) is stable if, for each time t_0, and for every constant $B > 0$, there exists some $b(B, t_0) > 0$ such that

$$\|x(t_0)\| < b \Rightarrow \|x(t)\| < B \ \forall t \geq 0 \tag{2.77}$$

Here

- $\|\bullet\|$ can be any vector norm.
- it is uniformly stable if b is independent of t_0.
- the equilibrium is unstable.
- equilibrium points of stable autonomous systems are necessarily uniformly stable, due to the state trajectories, of an autonomous system depend only on initial condition $x(t_0)$ and not on initial time t_0 (Figure 2.6).

Remarks

i. the solutions of (2.76a) are required to be defined for everything.
ii. such a condition of the global existence of a solution is not guaranteed by the local condition of Lipschitz.
iii. Lyapunov's theorem will ensure the existence of a global solution.
iv. in 1892 Lyapunov showed that a certain function can be used to determine the stability of a point of equilibrium [25, 26].

2.4.1.2   Asymptotic Stability
The equilibrium $x = 0$ of (2.76a) is asymptotically stable if:

(a) it is stable, and
(b) for each time t_0 there exists some $b(t_0) > 0$ such that

$$\|x(t_0)\| < b \Rightarrow \|x(t)\| \to 0 \ t \to \infty \tag{2.78}$$

Figure 2.6   Geometry stability concept.

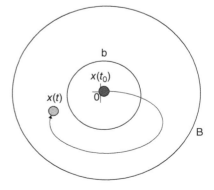

Remark

- It is uniformly asymptotically stable if it is asymptotically stable and both b and the rates of convergence in (b) are independent of t_0.
- Notice that stability of a non-trivial solution of a differential equation does not imply that the solution is bounded. Also, the boundedness of a solution of a differential equation does not imply that the solutions are stable. The concepts of boundedness and stability of solutions are mutually independent [25, 26].

2.4.1.3 Exponential Stability

The equilibrium of (2.76a) is exponentially stable if constants $b, B, > 0$ exist such that

$$\|x(t_0)\| < b \Rightarrow \|x(t)\| \leq Be^{-at}\forall t \geq t_0 \tag{2.79}$$

Remarks

- Asymptotic and exponential stability are local properties of a dynamic system since they only require that the state converges to zero from a finite set of initial conditions (known as a region of attraction):

$$\|x\| < b$$

- A strictly stable linear system is necessarily globally exponentially stable [25, 26].

2.4.2 The Direct Method of Lyapunov (Second Method)

2.4.2.1 Positive Function

The aim of Lyapunov's direct method is to determine the stability properties of an equilibrium point of an unforced nonlinear system without solving the differential equations describing the system. The basic approach involves constructing a scalar function V of the system state x, and considering the derivative $\dot{V}$ of V with respect to time. If V is positive everywhere except at the equilibrium $x = 0$, and if furthermore $\dot{V} \leq 0$ for all x (so that V cannot increase along the system trajectories), then it is possible to show that $x = 0$ is a stable equilibrium. This is the main result of Lyapunov's direct method, and a function V with these properties is known as a Lyapunov function.

Definition 2.11

Autonomous Systems

- A continuous function with this property is said to be **positive definite**. Alternatively, if $V(x)$ is continuous and $V(x) > 0 \, x \neq 0$, $V(x) = 0 \, x = 0$ holds for all x such that $\|x\| < B_0$ for some $B_0 > 0$, then V is locally positive definite or positive definite for $\|x\| < B_0$.
- Derivative of $V(x)$ along system trajectories. If x satisfies the differential equation $\dot{x} = f(x)$, then the time-derivative of a continuously differentiable scalar function $V(x)$ is given by

$$\dot{V}(x) = \nabla V(x)\dot{x} = \nabla V(x)f(x) \tag{2.80}$$

where $\nabla V(x)$ is the gradient of V (expressed as a row vector) with respect to x evaluated at x. The expression (2.80) gives the rate of change of V as x moves along a trajectory of the system state and $\dot{V}$ is therefore known as the derivative of V along system trajectories [25, 26].

Table 2.1 Review of the basic theorem of Lyapunov [27].

Definition	Basic conditions on $V(x, t)$	Basic conditions on $V(x, t)$	Remarks
1	LPDF	>0 locally	Stable
2	LPDF decrescent	≥ 0 locally	Uniformly stable
3	LPDF decrescent	LPDF	Asymptotically stable
4	PDF decrescent	PDF	Globally uniformly asymptotically stable

Non-Autonomous Systems

- The discussion so far has concerned only autonomous systems (systems with dynamics of the form $\dot{x} = f(x)$). Extensions to non-autonomous systems $\dot{x} = f(x, t)$ are straightforward, but the situation is complicated by the fact that a Lyapunov function for non-autonomous dynamics may need to be time-varying, i.e. of the form $V(x, t)$.
- Positive definite time-varying functions. To extend the definition of a positive definite function to the case of time-varying functions, we simply state that $V(x, t)$ is positive definite if there exists a time-invariant positive definite function $V_0(x)$ satisfying

$$V(x, t) \geq V_0(x), \forall t \geq t_0, \forall x \qquad (2.81)$$

- Decrescent functions. Non-autonomous systems differ from autonomous systems in that the stability of equilibrium may be uniform or non-uniform. In order to ensure uniform stability, a Lyapunov function V for a non-autonomous system must also be decrescent, which requires that

$$V(x, t) \leq V_0(x), \forall t \geq t_0, \forall x \qquad (2.82)$$

- Derivative of $V(x, t)$ along system trajectories. If $x(t)$ satisfies $\dot{x} = f(x, t)$, then the derivative with respect to t of a continuously differentiable function $V(x, t)$ can be expressed. The appearance here of the partial derivative of V with respect to t is due to the explicit dependence of V on time (Table 2.1)

$$\dot{V}(x, t) = \frac{\partial V}{\partial t}(x, t) + \nabla V(x)\dot{x} = \frac{\partial V}{\partial t}(x, t) + \nabla V(x)f(x, t) \qquad (2.83)$$

2.4.2.2 Theorem of Lyapunov
Stability/Asymptotic Stability for Autonomous Systems

Theorem 2.19 If there exists a continuously differentiable scalar function $V(x)$ such that:

i. $V(x)$ is positive definite.
ii. $V(x) = 0$, for all x satisfying $\|x\| < B_0$ for some constant $B_0 > 0$, then the equilibrium $x = 0$ is stable.
iii. $V(x)$ is positive definite whenever $\|x\| < B_0$, then $x = 0$ is asymptotically stable (Figure 2.7).

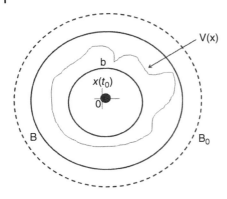

Figure 2.7   Stability Lyapunov.

Remarks

- The first part of this theorem can be proved by showing that it is always possible to find a positive scalar b which ensures that, for any given $B_0 > 0$, $x(t)$ is bounded by $\|x\| < B$ for all $t \geq t_0$ whenever the initial condition satisfies $\|x(t)\| < b$.
- Suppose that the origin is not asymptotically stable, then for any given initial condition $\|x(t_0)\|$ the value of $\|x(t)\|$ must remain larger than some positive number $B_0 < \|x(t_0)\|$ [25–27].

Stability/Asymptotic Stability for Non-Autonomous Systems

Theorem 2.20   If there exists a continuously differentiable scalar function $V(x, t)$ such that:

i. $V(x, t)$ is positive definite,
ii. $V(x) = 0$ for all x satisfying $\|x\| < B_0$ for some constant $B_0 > 0$, then the equilibrium $x = 0$ is stable: if, furthermore,
iii. $V(x, t)$ is decrescent then $x = 0$ is uniformly stable: if, in addition to (i), (ii), and (iii),
iv. $V(x, t)$ is positive definite whenever $\|x\| < B_0$, then $x = 0$ is uniformly asymptotically stable (see Table 2.2).

2.4.2.3   Globally Uniformly Asymptotically Stable of Lyapunov

Theorem 2.21   If V is a Lyapunov function for an autonomous system (or non-autonomous system) which satisfies conditions (i)–(iii) for all x, and V is radially unbounded, then the system is globally uniformly asymptotically stable [25–31].

2.4.2.4   Definition Matrices and Functions
Lyapunov functions are tools that characterize the various types of stability analysis of dynamical processes, i.e. they give information about the behavior of solutions in functions at initial conditions. Inspired by these properties of matrices and functions, researchers have been investigating the potential for applications for Lyapunov functions on the control of bioprocesses (see Table 2.2).

Table **2.2** Properties for the Lyapunov theorem.

Definition	Real functions $Q(x): \Re^n \to \Re$, $Q(x)=x^T Q x$		Square matrices $Q: n \times n$		Real quadratic functions, square symmetric matrix $Q = Q^T$, Eigenvalues
	Basic conditions on $Q(x)$	if	Basic conditions on Q	if	
Positive semi-definite	≥ 0	$Q(x) \geq 0, \forall x$	≥ 0	$x^T Q x \geq 0, \forall x$	$\lambda_Q \geq 0$
Positive definite	> 0	$Q(x) \geq 0, \forall x$; $Q(x) = 0$, if only if $x = 0$	> 0	$x^T Q x \geq 0, \forall x$; $x^T Q x = 0$, if only if $x = 0$	$\lambda_Q > 0$
Negative semi-definite	≤ 0	$Q(x) \leq 0, \forall x$	≤ 0	$x^T Q x \leq 0, \forall x$	$\lambda_Q \leq 0$
Negative definite	< 0	$Q(x) \leq 0, \forall x$; $Q(x) = 0$, if only if $x = 0$	< 0	$x^T Q x \leq 0, \forall x$; $x^T Q x = 0$, if only if $x = 0$	$\lambda_Q < 0$

2.4.3 The Indirect Method of Lyapunov (First Method)

The method is referred to as Lyapunov's indirect method or Lyapunov's first method. For the Lyapunov indirect method, system linearization around a given point (equilibrium-stationary) is used and one can converge to local stability on a small stability region.

2.4.3.1 Linearization

In many cases it is possible to determine whether the equilibrium of a nonlinear system is locally stable simply by examining the stability of the linear approximation to the nonlinear dynamics about the equilibrium point about the equilibrium $x = 0$ is derived from Taylor's series expansion of $f : D \rightarrow \mathfrak{R}^n$ about $x = 0$ where it is continuously differentiable and is a neighborhood of origin. Provided $f(x)$ is continuously differentiable, the linearization of a system (2.76a) derived from Taylor's series expansion of about $x = 0$ is

$$\dot{x} = Ax + \widetilde{f}(x) \qquad (2.84)$$

here satisfies

(i) $A = \left[\dfrac{\partial f}{\partial x}\right]_{x=0} = \begin{bmatrix} \dfrac{\partial f_1}{\partial x_1} & \dfrac{\partial f_1}{\partial x_2} \cdots & \dfrac{\partial f_1}{\partial x_n} \\ \vdots & \vdots & \vdots \\ \dfrac{\partial f_n}{\partial x_1} & \dfrac{\partial f_n}{\partial x_2} \cdots & \dfrac{\partial f_n}{\partial x_n} \end{bmatrix}_{x=0}$ A is the Jacobian matrix of f evaluated at $x = 0$

(ii) $\lim_{x \to 0} \dfrac{\|\widetilde{f}(x)\|_2}{\|x\|_2} = 0 \quad \forall t \geq 0$

Neglecting higher order terms gives the linearization of (2.76a)

$$\dot{x} = Ax \qquad (2.85a)$$

or

$$\ddot{x}(t) = Ax(t) - B\mu(t) \qquad (2.85b)$$

where $A = \frac{\partial f}{\partial x}(x_{eq}, u_{eq})$ and $B = \frac{\partial f}{\partial \mu}(x_{eq}, u_{eq})$. Matrices A and B are called the Jacobians. To summarize, Eq. (2.85b) represents the linearized dynamics of the nonlinear dynamic system (2.76a), around the equilibrium point (x_{eq}, u_{eq}). Very often, the linearized dynamics are sufficient to study energy systems, particularly around some desired operating point [25–27].

2.4.3.2 Stability by Linearization

Theorem 2.22 For the nonlinear system (2.84), suppose that f is continuously differentiable and define A as in (2.85a). Then:

i. $x = 0$ is an exponentially stable equilibrium of (2.84) if all eigenvalues of A have negative real parts
ii. $x = 0$ is an unstable equilibrium of (2.84) if A has at least one eigenvalue with a positive real part.

Stability for Linear Systems with Constant Coefficients We notice that any linear system has $f(x_{eq}, t) \equiv 0$ as an equilibrium solution .

If, in addition, the matrix of the system is non-singular, then this is the only equilibrium. We say that a linear system is stable or asymptotically stable or, respectively, unstable, when its equilibrium solution x_{eq} is stable or asymptotically stable ounstable.

Case 1D

We consider $\lambda \in \mathfrak{R}$ and the differential equation

$$\dot{x} = \lambda x \tag{2.86}$$

Theorem 2.23   If $\lambda < 0$ then Eq. (2.86) is asymptotically stable.
If $\lambda = 0$ then Eq. (2.86) is stable, but it is not asymptotically stable.
If $\lambda > 0$ then Eq. (2.86) is unstable.

Case 2D

We consider $A \in \mathfrak{R}^2$ with the eigenvalues $\lambda_1, \lambda_2 \in C$ and the system

$$\dot{x} = Ax \tag{2.87}$$

Theorem 2.24   If $\lambda_1 < 0$ and $\lambda_2 < 0$ then system (2.87) is asymptotically stable.
If $\lambda_1 < 0$ and $\lambda_2 < 0$ then system (2.87) is stable, but it is not asymptotically stable.
If $\lambda_1 < 0 \ \lambda_1 < 0$ or then system (2.87) is unstable [28].

2.4.4   Lasalles Invariance Principle

Theorem 2.25   If there exists a continuously differentiable scalar function $V(x)$ such that:

(i) $V(0) = 0$
(ii) $V(x) > 0$ for all $x \neq 0$
(iii) $\dot{V}(x) \leq 0$ for all x
(iv) $V(x) \rightarrow \infty$ *as* $\|x\| \rightarrow \infty$
(v) The only solution of $\dot{x} = f(x)$ such that $V(x) = 0$ is $x(t) = 0$ for all t

then $x = 0$ is globally asymptotically stable.

2.4.5   Invariant Set

Set $\wp$ is an invariant set for a dynamic system if every system trajectory starting in $\wp$ remains in $\wp$ at all future times.

Theorem 2.26   Let f in $\dot{x} = f(x)$ be continuous and assume that there exists a continuously differentiable scalar function $V(x)$ such that:

i. $V(x)$ is positive definite and radially unbounded.
ii. $\dot{V}(x) \leq 0$ then all solutions $x(t)$ of $\dot{x} = f(x)$ are globally bounded. Furthermore, let B be the set of all x for which $\dot{V}(x) = 0$, and let $\wp$ be the largest invariant set in B. Then every state trajectory $x(t)$ of $\dot{x} = f(x)$ converges to $\wp$ as $t \rightarrow \infty$.

Remarks

i. The set $\wp$ is defined in the theorem as the largest invariant set in B, in the sense that $\wp$ is the union of all invariant sets (e.g. equilibrium points or limit cycles) within B [25–27].

2.4.6 Input/Output Stability

The need to input-to-state stability (ISS) was introduced in [32, 33] and has proved to be a very useful paradigm in the study of nonlinear stability as well as its variants such as integral ISS and input/output stability (IOS) [34–40]. The definition of ISS (previously called simply detectability) takes into account the effect of initial states in a manner fully compatible with Lyapunov stability and incorporates naturally the idea of "nonlinear gain" functions [32, 33]. A system is ISS provided that, no matter what the initial state is, if the inputs are small, then the state must eventually be small. In order to input-output-to state stability (IOSS) formally, there is stability from the input/output data to the state, and IOS which refers instead to the stability of outputs [33, 37, 39]. By and large, the theory of state estimation is well understood for linear systems, but it is still poorly developed for more general classes of nonlinear systems. A still outstanding open question is the derivation of useful necessary and sufficient conditions for the existence of observers, i.e. "algorithms"(dynamical systems) which converge to an estimate $\hat{x}(t)$ of the state $x(t)$ of the system of interest, using the information provided by $\{u(s), s \le t\}$, the set of past input values, and by $\{y(s), s \le t\}$, the set of past output measurements.

However, the only way that this estimate fulfills the goal of upper bounding the norm of the true state as $t \to \infty$ is if $x(t) \to 0$. In other words, one obvious necessary property for the possibility of norm-estimation is that the origin must be a globally asymptotically stable state with respect to the "subsystem" consisting of those states for which the input $\mu \equiv 0$ produces the output $y \equiv 0$.

A system $\dot{x} = f(x, \mu)$ with measurement ("output") map $y = h(x)$ is IOSS if there are some functions $\rho \in KL$ and $\eta_1, \eta_2 \in K_\infty$ such that the estimate

$$|x(t)| \le \max\{\rho(|x(0)|, t), \eta_1(\|\mu_{[0,t]}\|), \eta_2(\|y_{[0,t]}\|)\} \tag{2.88}$$

- initial state $x(0)$
- input $\mu(\cdot)$
- $x(\cdot)$is the ensuing trajectory
- $y(t) = h(x(t))$
- $|\cdot|$to denote Euclidean norm
- $\|\cdot\|$ supremum norm

namely, there is a proper (radially unbounded) and positive definite smooth function V of states (a "storage function" in the language of dissipative systems introduced by Willems [40] and further developed by Hill and Moylan [41]) such that a dissipation inequality

$$\frac{d}{dt}V(x(t)) \le -\eta_1(|x(t)|) + \eta_2(|y(t)|) + \eta_3(|u(t)|) \tag{2.89}$$

This provides an "infinitesimal" description of IOSS, and a norm-observer is easily built from V [42]

2.4.7 General Properties of Linear and Nonlinear Systems

However, none of these properties of Linear Systems is, in general, true for nonlinear systems. The Lyapunov stability analysis of LTI systems is based entirely on quadratic forms, (see, Table 2.2) functions of the form

$$V(x) = x^T P x \tag{2.90}$$

where P is some positive definite symmetric matrix (Table 2.2). The derivative of V along trajectories of linear systems is

$$\dot{V}(x) = x^T (A^T P + PA)x \tag{2.91}$$

and the system is therefore globally asymptotically stable by Lyapunov's direct method if there exists a positive definite matrix Q satisfying the condition

$$A^T P + PA = -Q \tag{2.92}$$

Theorem 2.27 *Lyapunov Stability of LTI Systems*
A necessary and sufficient condition for an LTI system to be stable is that, for any positive definite symmetric matrix Q, the Lyapunov matrix Eq. (2.92) has a unique solution P which is a positive definite symmetric matrix.

Definition 2.12 *Passivity and Dissipativity*
The system (2.76a) is passive if there exists a continuous function $V(x, t)$ such that $V(0, t) = 0$ $V(x) \geq 0$, and along system trajectories

$$\dot{V}(x, t) = y^T(t)u(t) - g(t) \tag{2.93}$$

for all t, for some function $g \geq 0$.

2.4.8 Advanced Stability Theory

2.4.8.1 Concepts of Stability for Non-Autonomous Systems

The Lyapunov-like convergence theorems of this section contain the global asymptotic convergence results of Lyapunov's direct method as special cases; namely when $W(x)$ in the convergence theorem for non-autonomous systems is positive definite, and when M in the invariant set theorem consists only of the origin. A significant generalization of Lyapunov's direct method by providing criteria for convergence to entire state trajectories such as limit cycles.

The techniques for Lyapunov-like convergence analyses are slightly different for autonomous and non-autonomous dynamics. Considered the case of non-autonomous systems of the form

$$\dot{x} = f(x, t) f(0, t) = 0 \; \forall t \tag{2.94}$$

The method for autonomous systems, though more powerful and closer in principle to Lyapunov's direct method, will then be treated as a special case as for non-autonomous systems. Suppose that we have found a positive definite function $V(x, t)$ whose derivative

$\dot{V}$ along trajectories of (2.94) satisfies

$$\dot{V}(x, t) \le -W(x) \le 0 \qquad (2.95)$$

for some non-negative function W. The aim of the convergence analysis is to show that $W(x, t)$ tends to zero, and therefore that $x(t)$ converges to the set of states on which $W(x) = 0$.

Barbalat's lemma For any function $\chi(t)$, if $\chi(t)$ exists and is finite for all t $\lim_{t \to \infty} \int_0^t \chi(s)ds$ exists and is finite then $\lim_{t \to \infty} \chi(t) = 0$.

2.4.8.2 Lyapunov-like Analysis Using Barbalat's Lemma

Theorem 2.28 *Convergence for Non-Autonomous Systems*
Let the function f in (2.94) be continuous with respect to x and t, and assume that there exists a continuously differentiable scalar function $V(x, t)$ such that:

  i. $V(x, t)$ is positive definite, radially unbounded and decrescent.
 ii. $\dot{V}(x, t) \le -W(x) \le 0$.
    where W is a continuous function. Then all state trajectories $x(t)$ of (2.94) are globally bounded, and satisfy $W(x, t) \to 0$ as $t \to \infty$ [25–27].

2.5 Bifurcation Theory

The main purpose of bifurcation analysis is to characterize changes in the qualitative behavior of a mathematical model defined by ODE's and/or partial differential equations (PDE) by varying key parameters. For example, in the case of bioreactor operation it is common to use the dilution rate as a parameter of bifurcation. Many important features, as well as potential limitations hard to find with simple simulations, are uncovered through this analysis. Bifurcation analysis is applied to linear and nonlinear (bio) mathematical models; it can be used to find optimal operating conditions and to avoid hazardous conditions for the process [43–45]. This analysis is very useful and works by revealing all possible equilibrium points for a given initial operation condition, maintenance free one or two system parameters in many cases (bifurcation parameter) such as the dilution rate and initial substrate concentration. With the above analysis is possible to predict if the system can present oscillations, hysteresis, steady-state multiplicity and chaos [46–48].

The local bifurcation is determined by calculating the system eigenvalues as in the situation of stability analysis, it occurs when one of the eigenvalues is close to the axis of imaginary numbers in the complex plane. The simplest bifurcations are associated when one of the eigenvalues takes the value zero (fold bifurcation) as is the case of the branching point (BP) and limit point (LP) or when a pair of conjugate eigenvalues cross the imaginary axis (Hopf bifurcation, H) (Table 2.3).

The fold bifurcations are usually the cause of multiplicity of steady states and hysteresis. Hopf bifurcations are responsible for the appearance and disappearance of periodic solutions. As any bioreactor is not free of perturbations, it is important to have the information about the response of fermenter to those disturbances with the aim, in the worst case, of preventing the collapse of the bioprocess.

Table 2.3 Local stability R² [49].

Tr(A)	Det(A)	Tr(A)²-4Det(A)	Remarks	Stability
−	+	+	Real both (−)	Stable-node
−	+	−	Complex, real part (−)	Stable-focus
0	+	−	Imaginary, real part (−, 0)	Hopf bifurcation
+	+	−	Complex, real part (−)	Unstable-focus
+	+	+	Real both (−)	Unstable-node
±	0	+	One zero, one (−) o (+)	Saddle-node bifurcation
±	−	−	Real, one (−) o (+)	Saddle-point
0	0	0	Both zero	Double zero bifurcation point

Source: Reproduced with permission of John Wiley & Sons.

The bifurcation diagrams were obtained using the package Matcont 4.2 for MATLAB v7.0 or higher. This software is able to accomplish both steady state and dynamic bifurcation analyses including the determination of entire periodic solution branches using continuation techniques.

2.5.1 Periodic Orbit

A solution of $\dot{x} = f(x,,t), x \in R^n$ through the point x_0 is said to be periodic of period T if there exists $T > 0$ such that $x(t, t_0) = x(t + T, x_0)$ for all $t \in R$. Any periodic orbit in the phase space is a closed curve. The systems which can be written in the form $\dot{x} = -\nabla V$ for some continuously differentiable, single-valued scalar function $V(x)$ is called a gradient system with potential function V. Closed-orbit does not exist in gradient systems. If one can find a Lyapunov function for a system then also the existence of closed orbit can be ruled out [44].

2.5.2 Limit Cycle

A limit cycle is an isolated periodic orbit. The existence of a limit cycle in a model implies that the system exhibits self-sustained oscillations. For the existence of a limit cycle, dynamical system must be nonlinear and its phase space should be at least two dimensional. A linear system can have closed orbits but they are not isolated. The famous Poincare Bendixon theorem can help us to show the existence of limit cycles in two-dimensional phase space. An improved version of the Poincare Bendixon theorem states that if a trajectory is trapped in a compact region (a subset of two-dimensional phase space) then it must approach a fixed point, a limit cycle or something exotic called a cycle graph (an invariant set containing a finite number of fixed pints connected by a finite number of trajectories, all oriented either clockwise or counterclockwise) [50].

2.5.3 Bifurcation of Maps

If an eigenvalue reaches the unit circle, then the fixed point is no longer hyperbolic and a bifurcation can occur. The loss of hyperbolicity of a fixed point occurs in one of the three following ways:

i. A Jacobian matrix has a simple real eigenvalue 1. This type of bifurcation is known as steady-state bifurcation for maps. The saddle-node, transcritical, and pitchfork bifurcations are examples of steady-state bifurcation.

ii. A simple conjugate pair of eigenvalues of the Jacobian matrix lying on the unit circle. We refer to this case as Hopf bifurcation for maps.

iii. A simple real eigenvalue of the Jacobian matrix is −1. In this case, the period doubling bifurcation occurs. This bifurcation is also known as flip bifurcation or subharmonic bifurcation [51].

2.5.4 Hyperbolic and Non-Hyperbolic Equilibrium Points

A hyperbolic equilibrium point of a continuous dynamical system are those equilibrium points at which the Jacobian matrix has no eigenvalue with zero real part. On the other hand, if an equilibrium point is not hyperbolic then it is called non-hyperbolic. Hartman–Grobman theorem states that the stability type of the hyperbolic equilibrium point is fully captured by the linearized system. A phase portrait is structurally stable at hyperbolic equilibrium point i.e. topology of the phase portrait cannot be changed by an arbitrarily small perturbation to the vector field.

2.5.5 Bifurcation Point

A point (x_0, π) is called bifurcation point of a dynamical system (with parameter π) if systems behavior changes qualitatively at (x_0, π). A necessary condition for (x_0, π) to be a bifurcation point is that it must be a nonhyperbolic equilibrium point at the critical parameter value. Hartman–Grobman theorem implies that any qualitative change or bifurcation must be reflected in the linear dynamics. The qualitative structures of equilibrium point remain fixed unless the equilibrium point loses its hyperbolicity. The loss of hyperbolicity of equilibrium point occurs in one of the two following ways.

i. A simple real zero eigenvalue of the Jacobian matrix at the critical parameter value. This type of bifurcation is called steady state bifurcation. Most typical steady-state bifurcations are a saddle-node bifurcation, transcritical bifurcation, and pitchfork bifurcation.

ii. A simple pair of purely imaginary eigenvalues of the Jacobian matrix at the critical parameter value. This type of bifurcation is known as Hopf bifurcation [52].

2.5.6 Lyapunov Exponent

The Lyapunov exponent or Lyapunov characteristic exponent of a dynamical system is a quantity that characterizes the rate of separation of infinitesimally close trajectories with time. Quantitatively, two trajectories in phase space with initial separation dZ_0 changes with time using the following rule.

$$|dZ(t)| \approx e^{\lambda t}|dZ_0| \tag{2.96}$$

where λ is the Lyapunov exponent.

The rate of separation can be different for different orientations of the initial separation vector. An n dimensional dynamical system has n Lyapunov exponents.

i. For a periodic behavior of a system largest Lyapunov exponent should be negative.
ii. For a quasiperiodic behavior, the largest Lyapunov exponent should be 0.
iii. For chaotic behavior, the largest Lyapunov exponent must be positive.
iv. For a hyperchaotic system at least two Lyapunov exponents are positive [53].

2.5.7 Chaos

A dynamical system is said to be chaotic if it has the following properties:

i. it must be sensitive to initial conditions.
ii. it must be topologically mixing.
iii. its periodic orbits must be dense.

The necessary requirements for a deterministic continuous dynamical system to be chaotic are that the system must be nonlinear and be at least three-dimensional. At least one Lyapunov exponent must be positive for chaotic systems. For a hyper-chaotic system at least two Lyapunov exponents are positive. Sensitivity to initial conditions means that each point in such a system is arbitrarily closely approximated by other points with significantly different future trajectories.

Topologically mixing means that the system will evolve over time so that any given region or open set of its phase space will eventually overlap with any other given region.

Linear systems are never chaotic; for a dynamical system to display chaotic behavior, it has to be nonlinear. Also, by the Poincare Bendixon theorem, a continuous dynamical system on the plane cannot be chaotic; among continuous systems, only those whose phase space is non-planar (having at least three dimensions) can exhibit chaotic behavior. However, a discrete dynamical system (such as the logistic map) can exhibit chaotic behavior in one-dimensional or two-dimensional phase space [54].

2.5.8 Topological Equivalence

Theorem 2.29 *One-Dimensional Equivalence*
Two flows are topologically equivalent if their equilibria, ordered on the line, can be put into one-to-one correspondence and have the same topological type (sink, source or semi-stable). Topological conjugacy requires more than that the trajectories of one map continuously onto those of the other. It also requires that the temporal parameterization be the same.

Two flows are topologically equivalent if there exists a homeomorphism that maps the orbits of one onto the orbits of the other and preserves the direction of time [55]. That is, look for an increasing map such that the flows $\varphi_t : A \to A$ and $\phi_t : B \to B$ are topologically conjugate if there is a homeomorphism $h : A \to B$ such that

$$h(\varphi_t(x)) = \phi_t(h(x)) \tag{2.97}$$

2.5.9 Example Bifurcations and Structural Stability of Dynamical Systems

A large number of applications are modeled by autonomous systems of differential equations that contain parameters. As these change, the stylized phase images of differential equations can also be altered; the values of the parameters in which these modifications are presented are called bifurcation values. In the most general sense, bifurcation in a family of dynamic systems means a qualitative change in the stylized phase images when a parameter reaches or exceeds a certain critical value. More specifically, we call the bifurcation of a point of equilibrium (or invariant set) the case where this change consists of the appearance of new points of equilibrium (or invariant sets) that move away from the point or set in order to move away from the critical value parameter. The bifurcation does not depend on the properties of the linear parts of the systems, but it can occur even when they do not change. The true root of very broad classes of bifurcation must be sought in certain principles of dynamic persistence [56–58].

Mathematical models for biological oscillators accommodate this nonlinear kinetics either explicitly, or phenomenological (through Michaelis–Menten or Hill functions). The latter is often privileged because detailed molecular mechanisms are generally not known and because it greatly simplifies the models. In the example in our work, the importance of a characteristic feature of the Goodwin and related models is that degradation reactions, clock mRNA, and clock protein plays an important role in the control of the oscillator's period. The model involves a single gene. The mRNA, X, is translated into enzyme Y, which catalyzes the production of metabolite Z, which causes inhibition of expression (by activating an unmodeled repressor). Neglecting the specifics of catalysis and inhibition, Goodwin formulated the model in terms of concentrations X, Y, and Z as:

$$\frac{dX}{dt} = \frac{a}{k^n + Z^n} - bX \tag{2.98}$$

$$\frac{dY}{dt} = \alpha X - bY \tag{2.99}$$

$$\frac{dZ}{dt} = \gamma Y - \delta Z \tag{2.100}$$

The mRNA concentration rises, followed by a rise in enzyme concentration, and then a rise in metabolite concentration. The rise in Z causes a crash in X, which produces a metabolite (Z), which (indirectly) represses gene expression. This negative feedback, coupled with the delay inherent in the three-step loop, can result in oscillatory behavior (Table 2.4).

According to the experimental exercise that we were given, which was previously evaluated with a Goodwin oscillator model, we identified certain parameters such as: translation rate, transcription rate, the constant rate for metabolite degradation in addition to the rate of catalysis (Figures 2.8 and 2.9). These parameters gave us the clue to think that it could be a reactor in which a process has been established in which we are interested in getting a certain product of commercial interest, however it is fundamental to evaluate each of the parameters that govern our system of equations previously established, this in order to

Table 2.4 Parameters used in modeling and their meaning.

Parameters	Symbol	Meaning	Value
Alfa	α	The translation rate	1
Union constant	k	The constant of the union of the final product to the transcription factor	1.368
Hill coefficient	n	Indicates how many substrate binding zones of an enzyme affect the affinity of the substrate binding in the rest of the binding zones	12
Degradation speed	b	Velocity constant for mRNA degradation	1
Transcription rate	a	Transcription rate	360
Beta	β	Velocity constant for protein degradation	0.6
Gamma	γ	Catalysis rate	1

Figure 2.8 The Goodwin oscillator. Phase portrait showing convergence to a limit cycle in the three-dimensional phase space.

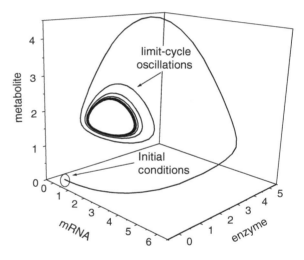

identify our bifurcation values and at the same time be able to give our system maximum stability, in order to achieve an efficient process [59].

2.6 Overview of Non-Smooth Dynamical Systems

Non-smooth dynamical systems are also known to exhibit sudden losses of structural stability under parameter variations, in concrete problems in applied science and engineering where non-smooth phenomena play an important role [60, 61]. They are also present in system networks if non-smooth characteristics are used to represent switches and in

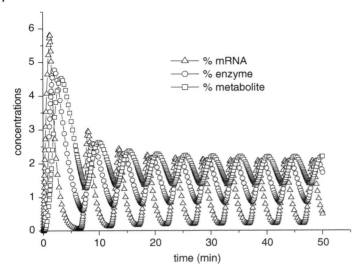

Figure 2.9 Dynamics evolution of the concentrations described by Eqs. (2.98)–(2.100).

control theory, they frequently appear when discontinuous controls are involved. The visible equilibria, invisible equilibria, and pseudo-equilibrium on the switching manifold are studied with the application of the Filippov method [61, 62]. The global bifurcation of periodic solutions is also investigated due to the interaction of system portraits with switching manifold. Following [63] we divide the non-smooth dynamical systems of interest into three different categories considering the system (2.76a), depending on the discontinuity type of their orbits and vector fields.

i. Non-smooth systems whose orbits and vector fields are everywhere continuous, or piecewise smooth continuous systems. Such systems have continuously-differentiable orbits but discontinuities in the first or higher derivatives of f.

ii. Systems exposed to discontinuities (or jumps) in the state or impacting systems. Such systems are more commonly formulated as a hybrid system with discontinuous jumps described by auxiliary maps.

iii. Systems with discontinuous vector fields (Filippov systems).

In general, mathematical description of the planar non-smooth single-input single-output (SISO) system is as follows

$$\dot{X} = \begin{cases} F(X) & h(X) \leq 0 \\ F(X) + \mu G(X) & h(X) > 0 \end{cases} \qquad (2.101)$$

with μ control parameter and $X \in \mathfrak{R}^2$ non-smooth system (2.101) governed by two different smooth ODEs which evolve in their own subspace of phase portraits. The so-called switching manifold is the boundary or smooth section between two phase subspaces. The system (2.101) is a SISO system with scalar $H(X)$ due to quasi identically transformation method [64–73].

List of Figures

List of Tables

References

1 Poria, S. and Dhiman, A. (2016). *Existence and Uniqueness Theorem for ODE: An Overview*. India: Department of Applied Mathematics, University of Calcutta.

2 Aström, K.J. and Murray, R.M. (2009). *Feedback Systems: An Introduction for Scientists and Engineers*. New Jersey, EEUU: Princeton University Press.

3 Gomez-Acata, R.V., Pablo, A., Lopez-Perez, P.A. et al. (2012). Bifurcation analysis of continuous aerobic nonisothermal bioreactor for wastewater treatment. *IFAC Proceedings Volumes* 45 (12): 24–29.

4 Aguilar-López, R., López-Pérez, P.A., Neria-González, M.I., and Domínguez-Bocanegra, A.R. (2010). Modelo adaptable basado en un observador para una clase de bio-reactor aerobio por lotes. *Revista mexicana de ingeniería química* 9: 29–35.

5 Maqueda, M.A.M., Martinez, S.A.D., Narváez, D. et al. (2006). Dynamical modelling of an activated sludge system of a petrochemical plant operating at high temperatures. *Water Science and Technology* 53 (11): 135–142.

6 Aström, K. J. and Murray, R. M. (2009) *Feedback Principles: System Modeling* Chapter 2, 31-60 S.

7 Hairer, E., Norsett, S.P., and Wanner, G. (1993). *Solving Ordinary Differential Equations I: Nonstiff Problems*, 2e. Berlin: Springer-Verlag.

8 Rowell, D. (2002). *Analysis and Design of Feedback Control Systems State-Space Representation of LTI Systems*. MIT.

9 Inman, D.J. and Andry, A.N. Jr., (1980). Some results on the nature of eigenvalues of discrete damped linear systems. *ASME Journal of Applied Mechanics* 47 (4): 937–930.

10 Seyranian, A.P. (1993). Sensitivity analysis of multiple eigenvalues. *Mechanics of Structures and Machines* 21 (2).

11 Mukherjee, N. and Poria, S. (2012). Preliminary concepts of dynamical. *International Journal of Applied Mathematical Research* 1 (4): 751–770.

12 Cid, J.A. and Pouso, R.L. (2009). Does Lipschitz w.r.t x imply uniqueness for $\mathbf{y}'=(\mathbf{x},\mathbf{y})$? *The Mathematical Association of America* 116.

13 Shadid, J.N., Tuminaro, R.S., and Walker, H.F. (1997). An inexact Newton method for fully coupled solution of the Navier–stokes equations with heat and mass transport. *Journal of Computational Physics* 137: 155–185.

14 Aguilar, A. P. (2012) Nonlinear Control Systems *Fundamental properties IST-DEEC PhD Course*

15 Megretski, A. (2003). *Dynamics of Nonlinear Systems Institute of Technology*. Massachusetts: Department of Electrical Engineering and Computer Science.

16 Isidori, A. (1995). *Nonlinear Control Systems*. Secaucus, USA: Springer- Verlag, New York.

17 David M. Grobman (1959) Homeomorphism of systems of differential equations, *Nonlinear Dynamics And Chaos: With Applications To Physics, Biology, Chemistry*.

18 Sakka, A. (2001) Linear problems and hierarchies of Painlev'e equations, *Journal of Physics A: Mathematical Theory and Engineering*, Westview Press42 025210 1 19.

19 Cannon, M. (2009) Nonlinear Systems.

20 Slotine, J.-J. and Li, W. (1991). *Applied Nonlinear Control*. Prentice-Hall.

21 Vidyasagar, M. (1993). *Nonlinear Systems Analysis*. Prentice-Hall.

22 Khalil, H.K. (1996). *Nonlinear Systems*, 2e. Prentice-Hall.

23 Cannon, M. (2018). *Nonlinear Systems*. Michaelmas Term.

24 Cheng, G. (2004). *Stability of Non Linear Systems*, 4881–4896. New York: Wiley.

25 Murray, R. M., Li, Z. and Sastry, S. S. (1994) A Mathematical Introduction to Robotic Manipulation.

26 Chiang, H.D., Hirsch, M., and Wu, F. (1988). Stability regions of nonlinear autonomous dynamical systems. *IEEE Transactions on Automatic Control* 33 (1): 16–27.

27 Sassano, M. and Astolfi, A. (2013). Dynamic Lyapunov functions. *Automatica* 49: 1058–1067.

28 Bacciotti, A. (1992). *Local Stabilizability of Nonlinear Control Systems*. World Scientific.

29 Bacciotti, A. and Rosier, L. (2005). *Lyapunov Functions and Stability in Control Theory*. Springer-Verlag.

30 Sassano, M., & Astolfi, A. (2011). Dynamic Lyapunov functions. In Proc. of the 18th IFAC world congress, Milan

31 Krichman, M., Sontag, E.D., and Wang, Y. (2001). Input-output-to-state stability. *SIAM Journal on Control and Optimization* 39 (6): 1874–1928.

32 Sontag, E.D. (1989). Smooth stabilization implies coprime factorization. *IEEE Transactions on Automatic Control* 34: 435–443.

33 Christofides, P.D. and Teel, A. (1996). Singular perturbations and input-to-state stability. *IEEE Transactions on Automatic Control* 41: 1645–1650.

34 Jiang, Z.P., Teel, A., and Praly, L. (1994). Small-gain theorem for ISS systems and applications. *Mathematics of Control, Signals, and Systems* 7: 95–120.

35 Krstic, M., Kanellakopoulos, I., and Kokotovic, P.V. (1995). *Nonlinear and Adaptive Control Design*. New York: Wiley.

36 Praly, L. and Wang, Y. (1996). Stabilization in spite of matched unmodelled dynamics and an equivalent definition of input-to-state stability. *Mathematics of Control, Signals, and Systems* 9: 33.

37 Sontag's, J.T. (1993). Input to state stability condition and global stabilization using state detection. *Systems and Control Letters* 20: 219–226.

38 Sontag, E.D. and Wang, Y. (1999). Notions of input to output stability. *Systems and Control Letters* 38: 351–359.

39 Sontag, E.D. and Wang, Y. (2000). Lyapunov characterizations of input to output stability. *SIAM Journal on Control and Optimization* 39: 226–249.

40 Willems, J.C. (1976). Mechanisms for the stability and instability in feedback systems. *Proceedings of the IEEE* 64: 24–35.

41 Hill, D.J. and Moylan, P. (1992). Dissipative nonlinear systems: basic properties and stability analysis. In: *Proceedings of the 31st IEEE Conference on Decision and Control*, 3259–3264. IEEE.

42 Sontag, E.D. and Wang, Y. (1997). Output-to-state stability and detectability of nonlinear systems. *Systems and Control Letters* 29: 279–290.

43 Namjoshi, A., Kienle, A., and Ranmkrishna, D. (2003). Steady state multiplicity in bioreactors: bifurcation analysis of cybernetic models. *Chemical Engineering Science* 58: 793–800.

44 Gomez-Acata, R. V., Lopez-Perez, P. A., Maya-Yescas R. and Aguilar-Lopez, R. (2012) Dynamic Behavior Analysis of Carboxymethylcellulose Hydrolysis in a Chemostat. *Analysis and Control of Chaotic Systems*.

45 Gómez-Acata R.V., López-Pérez, P.A., Maya-Yescas et al. (2012). Bifurcation Analysis of Continuous Aerobic Nonisothermal Bioreactor for Wastewater Treatment. *Analysis and Control of Chaotic Systems-IFAC*. 45 (12): 24–29.

46 Femat, R., Méndez-Acosta, H.O., Steyer, J.P., and González-Alvarez, V. (2004). Temperature oscillations in abiological reactor with recycle. *Chaos, Solitons & Fractals* 19 (4): 875–889.

47 Alvarez-Ramirez, J., Alvarez, J., and Velasco, A. (2009). On the existence of sustained oscillations in a class of bioreactors. *Computers & Chemical Engineering* 33 (1): 4–9.

48 Zhang, Y. and Henson, M.A. (2001). Bifurcation analysis of continuous biochemical reactor models. *Biotechnology Progress* 17 (4): 647–660.

49 Gray, P. and Scoot, S.K. (1990). Chemical oscillations and instabilities. Non-linear chemical kinetic. Clarendon Press. Oxford.

50 Abed, E.H. and Fu, J.H. (1986). Local feedback stabilization and bifurcation control, I. Hopf bifurcation. *Systems and Control Letters* 7: 11–17.

51 Abed, E.H. and Wang, H.O. (1995). Feedback control of bifurcation and chaos in dynamical systems. In: *Nonlinear Dynamics and Stochastic Mechanics* (eds. W. Kliemann and N. Sri Namachchivaya), 153–173. Boca Raton, FL: CRC Press.

52 Van Opstale, M. (1998). Quantifying chaos in dynamical systems with lyapunov exponents. *Electronic Journal of Undergraduate Mathematics* 1: 1–8.

53 Mukherjee, N. and Poria, S. (2012). Preliminary concepts of dynamical systems. *International Journal of Applied Mathematical Research* 1 (4): 751–770.

54 Kybernetika (1995). Topological equivalence and topological linearization of controlled dynamical systems. *Sergej Čelikovský1* 31 (2): 141–150.

55 Chen, G., Moiola, J., and Wang, H. (2000). Bifurcation control: theories, methods, and applications. *International Journal of Bifurcation and Chaos* 10 (3): 511–548.

56 Woller, A., Gonze, D., and Erneux, T. (2013). The strong feedback limit of Goodwin circadian oscillator. *Physical Review E* 87: 032722.

57 Woller, A., Gonze, D., and Erneux, E. (2014). The Goodwin model revisited: Hopf bifurcation, limit-cycle, and periodic entrainment. *Physical Biology* 11 (4).

58 Goodwin, B.C. (1965). Oscillatory behaviour in enzymatic control processes. *Advances in Enzyme Regulation* 3: 425–428.

59 John, M., Thompson, T., and Giles, W. (1977). A bifurcation theory for the instabilities of optimization and design post Synthese. *Mathematical Methods of the Social Sciences Part II* 36 (3): 315–351.

60 Kowalczyk, P. (2006). Two parameter non-smooth bifurcations of limit cycles: classification and open problems. *International Journal of Bifurcation and Chaos* 16: 601–629.

61 Kuznetsov, Y.A., Rinaldi, S., and Gragnani, A. (2003). One parameter bifurcation in planar Filippov systems. *International Journal of Bifurcation and Chaos* 13: 2157–2188.

62 Suqi, M. (2017). Bifurcation of a non-smooth predator-prey system with prey-protected control strategy. *Journal of Physics A: Mathematical and Theoretical* 8: 4.

63 di Bernardo, M., di Bernardo, M., Champneys, A.R. et al. (2005, 2005). *Bifurcations and Chaos in Piecewise Smooth Systems; Theory and ApplicationBifurcations and Chaos in Piecewise Smooth Systems; Theory and Application* (ed. et al.). Springer-Verlag, Springer-Verlag.

64 di Bernardo, M., Garofalo, F., Iannell, L., and Vasca, F. (2003). Bifurcations in piecewise-smooth feedback systems. *International Journal of Control* 75: 1243–1259.

65 Hosham, H.A., Eman, D., and Elela, A. (2019). Discontinuous phenomena in bioreactor system. *Discrete and Continuous Dynamical Systems - B* 24 (6): 2955–2969.

66 Gao, Y., Meng, X., and Lu, Q. (2016). Border collision bifurcations in 3D piecewise smooth chaotic circuit. *Applied Mathematics and Mechanics* 37: 1239–1250.

67 Coombes, S. and Thul, R. (2016). Synchrony in networks of coupled non-smooth dynamical systems: extending the master stability function. *European Journal of Applied Mathematics* 27 (6): 904–922.

68 Yang, Q., Wang, L., Feng, E. et al. (2016). Identification and robustness analysis of nonlinear hybrid dynamical system of genetic regulation in continuous culture. *Journal of Industrial & Management Optimization* 13 (5): 1–21.

69 Wang, L., Yuan, J.L., Wu, C., and Wang, X. (2019). Practical algorithm for stochastic optimal control problem about microbial fermentation in batch culture. *Optimization Letters* 3 (3): 527–541.

70 Zifang, Q., Zhengdi, Z., Miao, P., and Qinsheng, B. (2018). Non-smooth bursting analysis of a Filippov-type system with multiple-frequency excitations. *Pramana* 91: 1–10.

71 Choudhury, S.R. and Mandragona, D. (2019). A chaotic chemical reactor with and without delay: competitive modes, and amplitude death bifurcations. *International Journal of Bifurcation and Chaos* 29: 02.

72 Ruks, L. and Van Gorder, R.A. (2017). On the inverse problem of competitive modes and the search for chaotic dynamics. *International Journal of Bifurcation and Chaos* 27 (1730032): 1–12.

73 Zhu, L. and Huang, X. (2010). Bifurcation in the stable manifold of a chemostat with general polynomial yields. *Mathematical Methods in the Applied Sciences* 33 (3): 340–349.

Part II

Observability and Control Concepts

3

State Estimation and Observers

From a bioprocess point of view, a wide variety of data cannot be directly obtained through measurement for the experimental systems. For economic or technological reasons, we are unable to place as many sensors as we want to measure the internal information since the cost is prohibitive, or it is sometimes not possible. In addition, more often than not, signals from sensors are distorted and tainted by measurement noises and some kinds of inputs and parameters are unknown or are not measured. Nowadays, in automatic control, the estimation covers at least the following: identification of uncertain parameters in the system equations, including delays, estimation of state variables, which are not measured, and observation fault diagnosis (unknown input) and isolation. With the available external measurement (output), is it possible to reconstruct the internal information (states)? This chapter introduces to the reader at the observability concepts that are the basis of the design of online estimation algorithms (software sensor) for bioprocess. The observability conditions for bioprocess models from local linearization, differential geometric, and algebraic differential approach are established.

3.1 Observability

3.1.1 Context and Motivations

In particular, for bioprocesses, their optimal performances depend on available information. Because these variables are frequently associated with process output quality, they are very important for process control and monitoring. For this reason, it is of great importance to deliver additional information about process variables, which is the role of the software sensor [1–3]. Given that some process variables are measured, parameter estimators and state observers (also called, "software sensors," state observers or simply observers) can be developed to estimate unknown parameters and to observe non-measured variables [4].

However, the development, and especially the implementation of advanced monitoring and control strategies of real bioprocesses, are difficult because of the absence of reliable instrumentation for the biological state variables, i.e. the substrates, biomass, and product concentrations. For example, the required quality of monitored data, precision data, time delay, frequency of sampling, are a function of the accuracy of bio-sensors, and usually require more sophisticated measurement devices. This can have several drawbacks, e.g.

Control in Bioprocessing: Modeling, Estimation and the Use of Soft Sensors, First Edition.
Pablo Antonio López Pérez, Ricardo Aguilar López, and Ricardo Femat.
© 2020 John Wiley & Sons Ltd. Published 2020 by John Wiley & Sons Ltd.

sterilization, discrete-time (and often rare) samples, relatively long processing (analysis) time, in many cases the state variables, are not online (and real-time), this is related to high-cost sensors and extreme operating conditions.

Such techniques have been successfully implemented in bioreactors in most biotechnological fields, including wastewater treatment and biomolecule production [5, 6]. One can distinguish two kinds of state observer: the exponential state observer and the asymptotic observer [7, 8]. Exponential state observers contain a tuning parameter for convergence, but are strongly dependent on the model quality and have low efficiency at low dilution rates. Asymptotic observers do not have any tuning parameter and convergence may be low at low dilution rates but are independent of the process kinetics, which is often poorly known. With respect to parameter estimation, two kinds of the estimator are considered: the observer-based estimator and the linear regression estimator [7].

In this chapter we focus on the dynamical case, namely, on the efficient development and implementation of state estimation schemes (estimators, "software sensors," state observers or simply observers), that can be used to design control and optimization strategies in chemical and biochemical processes.

Three important reasons justify resorting to an estimator:

- Knowledge acquisition and modeling. For example, when one must determine a variable, which cannot be measured directly or to estimate key process parameters.
- Process control. For example, when one wishes to control a variable not directly measured, or which is not readily available or that unmeasured variables intervene in the design of a control law.
- Supervision and diagnosis. For example, when it is necessary to supervise a microorganism concentration in a biofilm to avoid unstable operating conditions [8].

From a systematic point of view, the inputs of an estimator are the process inputs (i.e. control variables and measured disturbances) as well as measurements available on-line in the process, while the outputs of an estimator are the estimates of the unmeasured variables. This concept is shown in Figure 3.1.

For biological processes, another approach has been popularized by Bastin and Dochain [9]. It relies on the underlying structure of many bioprocess models that consist of two parts: (i) a linear part based on mass-balance considerations; and (ii) a nonlinear term which describes the biological reaction rates (kinetics). It should be noted that this latter term refers to the core of the biological activity and is often very sensitive to medium growth conditions; moreover, the associated kinetic parameters are not, in practice, known accurately [3].

The purpose of this section is to discuss some conditions required in the system for possible solutions to the above-mentioned observability issues. Such conditions above all correspond to what are usually called observability conditions. A question, fundamental to the analysis of bioprocesses, is whether the state of the system can be uniquely determined from its output data? Specifically, given the dynamic description of the system and the observation process, we can ask under what conditions can the initial state of the system be determined uniquely on the basis of the observed output on a given time interval? This problem is called the *inverse* or *observability* problem. The test of a system's observability is a necessary prerequisite to the estimation (predicted value) of states and parameters from

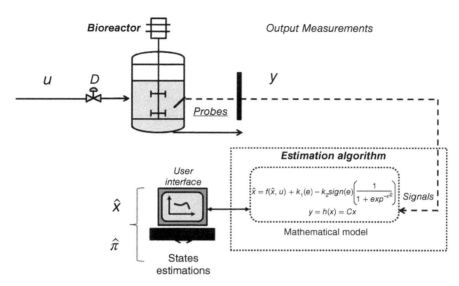

Figure 3.1   A general estimation scheme (x: state variables, π: parameters, y: outputs) [9].
Source: Reproduced with permission of Elsevier.

the output of the system. In short, they must express that there indeed is a possibility that the purpose of the observer can be achieved, namely that it might be possible to recover $x(t)$ on from the knowledge of u and y up to time t: at first glance, this will be possible only if $y(t)$ bears the information on the full state vector when considered over some time interval: this roughly corresponds to the notion of observability. Observability, in control theory, is a measure of how well internal states of a system can be inferred by knowledge of its external outputs. The observability and controllability of a system are mathematical duals.

Formally, a system is said to be observable if, for any possible sequence of state and control vectors, the current state can be determined in finite time using only the outputs (this definition is slanted toward the state space representation discussed in Chapter 2). Less formally, this means that from the system outputs it is possible to determine the behavior of the entire system (bioprocess). If a system is not observable, this means the current values of some of its states cannot be determined through output sensors: this implies that their value is unknown to the controller and, consequently, that it will be unable to fulfill the control specifications referred to by these outputs.

The class of systems under consideration is described by a state-space representation generally of the following form:

$$\dot{x}(t) = f(x(t), u(t))$$
$$y = h(x(t)) \tag{3.1}$$

where x denotes the state vector, taking values in X a connected manifold of dimension n, u denotes the vector of known external inputs, taking values in some open subset U, and y denotes the vector of measured outputs taking values in some open subset Y. Functions f and h will, in general, be assumed to be C^∞ of their arguments, and input functions $u(\circ)$ to be locally essentially bounded and measurable functions in a set U. The system will be assumed to be forward complete.

However, the preceding problem with regard to observer design is to analyze the observability conditions of the nonlinear systems under study. For linear systems, the classical observability index for the observability matrix, and the Observability Gramian [10] for the observability analysis and the estimator design, have been extensively studied, and have proved to be extremely useful, especially for on-line monitoring and control applications such as observer based control design. However, for nonlinear systems, the theory of observers is not nearly as complete or successful as it is for linear systems.

3.1.2 Linear Observability

As a background, consider the following linear system representation:

$$\dot{x} = Ax + Bu \tag{3.2}$$

$$y = Cx \tag{3.3}$$

where $x_0 \in \mathfrak{R}^n$ is the state vector $u \in \mathfrak{R}^p$ is the system input. A, B, C are the constant matrices with appropriate dimensions. $y \in \mathfrak{R}^q$ is the system output. The initial conditions for the systems are specified as

$$x(0) = x_0$$

System (3.2)–(3.3) is said to be observable if, at any time t the state $x(t)$ can be determined from the input sequence $u(s)$ and the output $y(s)$, $0 < s < t$. $x(t)$ can be determined if the initial state x_0 and the inputs are known. Hence, observability can equivalently be defined as the problem of finding the initial states from the given input–output measurements. If the initial states cannot be determined from the given sequence of inputs and output measurements, then the system is not observable.

For nonlinear systems a transformation to a linear system is necessary, this transformation is done via Taylor series linearization and it is valid for an equilibrium point and its neighborhood as given by the Grobman–Hartman theorem [10].

Here x is the variable state vector and y is the measured output vector. Therefore, by a Taylor series linearizing procedure:

$$\frac{dx}{dt} = A(\bar{x}) + B(\bar{u}) \tag{3.4}$$

where A, B, and x are:

$$A(\bar{x}) = \left[\frac{\partial f}{\partial x}\right]_{x=\bar{x}} \tag{3.5}$$

$$B(\bar{u}) = \left[\frac{\partial f}{\partial u}\right]_{u=\bar{u}} \tag{3.6}$$

Now, from a general framework, let us consider the following, let the finite sets be $Y = (y, y', y'', y'''...)^T \in \mathfrak{R}^m$ and $X = (x_1, x_2, x_3, x_4...)^T \in \mathfrak{R}^n$, with $m \le n$ which is related to the system output vector and finite time derivatives, it is therefore possible to construct the following linear dynamic system:

$$
\begin{bmatrix} y \\ y' \\ y'' \\ \vdots \\ y^{n-1} \end{bmatrix} = \begin{bmatrix} Cx \\ CAx \\ CA^2x \\ \vdots \\ CA^{n-1}x \end{bmatrix}
$$

or in vector form:

$$Y = NX \tag{3.7}$$

where

$$N = [C, CA, CA^2, CA^3, \dots, CA^{n-1}]^T \tag{3.8}$$

As can be seen, if the state vector can be determinate, the matrix N (called the observability matrix) must be invertible (full rank) in order to obtain:

$$X = N^{-1}Y \tag{3.9}$$

such that, the state vector X is observable with respect to the measurable output Y.

The local observability analysis is condensed as follows:

Local Observability Theorem. A continuous time linear (or linearized) system MA is *observable* if and only if

$$rank(N) = n, \tag{3.10}$$

where n is the order of the system [11].

Several unstable systems can generally be stabilized by using stabilizing control laws to place the system poles at desired locations. The inputs to the controller are the state measurements. In many systems, however, all the states may not be measurable. In that case, with the knowledge of the initial state, the state trajectory, can in principle be determined if (C, A) is observable. However, the procedure involves integration and inversion of a matrix which is ill-conditioned. As an alternative, more robust and practical approach, the states can be estimated using a state observer.

3.1.3 Nonlinear Observability

The design of observability conditions for nonlinear systems is a challenging problem (even for accurately known systems) that has received a considerable amount of attention. The first category of techniques consists of applying linear algorithms to the system linearized around the estimated trajectory. In a second approach, the nonlinear system is transformed into a linear one by an appropriate change of coordinates [12]. The corresponding analysis is computed in these new coordinates and the original coordinates are retrieved through the inverse transformation. In most approaches, nonlinear coordinate transformations are employed to transform the nonlinear system into block triangular observer canonical forms [13, 14]. Observability for a nonlinear system can be made locally effectively. But the observability criterion is similar to that of the linear case. The derivative of the output must be calculated (by using the Lie derivative), you must then build the observability matrix which requires that the rank of this matrix is full (n, the dimension the state of the system). In the linear case, you must derive not much more than $(n-1)$ times, because of the Cayley Hamilton Theorem. For the nonlinear case, unfortunately, this theorem does not exists. If it works up the order $(n-1)$, this is perfect; otherwise, you must compute the derivatives up to having a full order. In this section we present observability criteria for a class of nonlinear systems into the framework of differential geometry. Consider a single output nonlinear dynamical system in the following form:

$$\dot{x} = f(x)$$
$$y = h(x) \tag{3.11}$$

where $x \in \mathfrak{R}^n$ is the state vector and $y \in \mathfrak{R}^q$ is the system output. We assume that the vector field f and the output function h are smooth. In the following, we also assume that the pair $(f h)$ satisfies the observability rank condition. Thus, the so-called observability differential 1-forms are independent, and given by

$$\Theta_1 = d(h)$$
$$\Theta_i = dL^{i-1}_f(h) \ \forall \ 2 \le i \le n \tag{3.12}$$

where $L^k_f(h)$ is the kth Lie derivative of h along f and d is the differential operator.

3.1.4 Geometric Conditions of Observability

Let us to consider the following affine representation of the system (3.1):

$$\dot{x} = f(x) + g(x)u \tag{3.13}$$

with measured equation:

$$y = h(x) \tag{3.14}$$

under the assumption that $f(x) \in C^\infty$; $g(x) \in C^\infty$ and $h(x) \in C^\infty$.

Employing the Lie derivative of the measured output, the following is generated:

$$y = h(x)$$
$$\dot{y} = L_f h(x)$$
$$\vdots$$
$$y^{n-1} = L^{n-1}_f h(x)$$
$$y^n = L^{n-1}_f h(x) + L_g L^{n-1}_f h(x)u \tag{3.15}$$

where L^r_f is the r-order of the Lie derivative and $dL^r_f h$ are the Lie differential of r-order defined recursively as:

$$L^0_f h := h; \ dL^0_f h := dh = \left(\frac{\partial h}{\partial x_1}, \dots, \frac{\partial h}{\partial x_n} \right) \tag{3.16}$$

$$L^1_f h := \langle dh, f \rangle = \langle dh, f \rangle = \sum_{i=1}^n \frac{\partial h}{\partial x_1} f_i;$$

$$dL^1_f h := \left(\frac{\partial}{\partial x_1} \left(\sum_{i=1}^n \frac{\partial h}{\partial x_1} f_i \right), \dots, \frac{\partial}{\partial x_n} \left(\sum_{i=1}^n \frac{\partial h}{\partial x_1} f_i \right) \right) \tag{3.17}$$

$$L_f^r h := \langle dL_f^{r-1} h, f \rangle = L_f(L_f^{r-1} h), r \geq 2 \tag{3.18}$$

this defines a coordinate transformation $T(x)$ in order to represent the system (3.13) into a normal form, therefore:

$$\Theta = T(x) = \begin{bmatrix} \Theta_1 \\ \vdots \\ \Theta_n \end{bmatrix} = \begin{bmatrix} y = h(x) \\ \vdots \\ y^{n-1} = L_f^{n-1} h(x) \end{bmatrix} \tag{3.19}$$

Then a new transformed system can be defined via the new coordinates Θ as:

$$\Theta_1 = \Theta_2 = y$$

$$\vdots$$

$$\Theta_n = \Theta_{n-1} = y^n = L_f^n h(x) + L_g L_f^{n-1} h(x) u \tag{3.20}$$

The system represented by Eqs. (3.13) and (3.14) is at least locally uniformly observable for all, and the full rank observability condition is satisfied, as:

$$rank \left\{ \frac{\partial}{\partial x} K \right\} = n,$$

where K is the observability vector defined as

$$K = [dL_f^0 h, dL_f^1 h, \ldots, dL_f^{n-1} h]^T \tag{3.21}$$

3.1.4.1 Differential-Algebraic Observability Approach

Since the early 1990s, some papers have reviewed the dynamic characterization of a particular class of nonlinear systems named differentially flat [15] and Liouvillian systems [16], based on the frame of differential algebra. One of the most important uses of this approach for this kind of system is the explicit relationships that can be obtained for particular state variables; it is an advantage for a class of observation and control problems. Differential-algebra based techniques have been employed for differential algebraic as well as ordinary differential equations systems. Fliess and co-workers developed much of the work in the control literature on the employment of differential algebra. Important contributions to these techniques have been used for model identification [17].

In order to give the background to the estimation methodology proposed under the differential algebraic frame, the following definitions are considered:

Definition 3.1 Let L and K be differential fields. A differential field extension L/K is given by K and L such that: (i) K is a subfield of L and; (ii) the derivation of K is the restriction to K of the derivation of L.

Definition 3.2 An element of a differential field K is said to be *algebraic* if it satisfies a differential polynomial equation with coefficients over K. If an element is not algebraic then it is said to be *transcendent*.

Definition 3.3 Let $G, K < u >$ be differential fields. A *dynamic* consists of a finitely generated differential *algebraic* extension $G/K < u >.(G = K < u, \xi >, \xi \in G)$

Definition 3.4   An element in G is said to be *algebraically observable* with respect to $\{u, y\}$ if it is algebraic over $K < u, y>$.

So that a state x is said to be algebraically observable if it is algebraic over $K < u, y>$, that is to say, it satisfies a differential polynomial in terms of u, y and some of their time derivatives, i.e.

$$P(x, \ u, \ \dot{u}, \dots, y, \ \dot{y}, \dots) = 0 \tag{3.22}$$

with coefficients over $K < u, y>$. The above condition is called *the algebraic observability condition* (AOC).

3.1.5   Analytic Conditions for Observability

A fundamental question for the analysis of physical systems is whether the states of the system can be uniquely determined from its output data. Specifically, given the dynamic description of the system and the observation process, we can ask under what conditions can the initial state of the system be determined uniquely on the basis of the observed output on a given time interval? This problem is called the *inverse* or *observability* problem. The test of a system's observability is a necessary prerequisite to the estimation of states and parameters from the output of the system. In short, they must express that there is indeed a possibility that the purpose of the observer can be achieved, namely that it might be possible to recover x from only knowledge of u and y up to time t: at first glance, this will be possible only if the measured system output y bears the information on the full state vector when considered over some time interval: this roughly corresponds to the notion of observability.

3.1.6   Detectability

A linear system is detectable when all the unobservable modes are stable. This property is important for partially observable systems. An observable system is also detectable.

The property of detectability is important for control because one may successfully design a control system for an unobservable but detectable system so as to estimate and control the unstable modes.

A detectability analysis is done to determine if the unobservable subspace of the studied system is stable. When the systems have an unobservable subspace $(rank(N) < n)$, it is at least expected that the system will be stable in the equilibrium point. This analysis is done via the Popov–Belevitch–Hautus (PBH) theorem, which indicates that:

Local Detectability Theorem. A continuous time linear (or linearized) system CA is *detectable* if and only if *rank* $\begin{bmatrix} C - \lambda I \\ A \end{bmatrix} = n$ [11].

Where N contains the eigenvalues of the matrix C, I is the identity matrix.

3.1.7   Unobservable Subspaces

A state $x_0 \in \mathfrak{R}^n$ of the system (3.2) and (3.3) is said to be unobservable if

$$Ce^{At}x_0 = 0 \tag{3.23}$$

For all $t < 0$. That is, a state x_0 is unobservable if the zero-input response of the system (3.2) and (3.3) is zero for all time $t \geq 0$ whenever $x(0) = x_0$.

Definition 3.5 The unobservable Gramian of the system (3.2) and (3.3) is the matrix

$$W_{cn}(t_o, t_1) = \int_{t_0}^{t_1} e^{AT(\tau - t_0)} C^T C e^{A(\tau - t_0)} d\tau \in \mathfrak{R}^{nxn}, \quad t_o \leq t_1 \tag{3.24}$$

Clearly, we have that $W_0(t_o, t_1) \geq$ for all $t_o, t_1 \in \mathfrak{R}^n$ with $t_o \leq t_1$.

Lemma 3.1 A state $x_0 \in \mathfrak{R}^n$ is unobservable if and only if $x_0 \in N(W_0(0, T))^n$ for all $T > 0$.

3.1.7.1 A Geometric Characterization

Theorem 3.1 The unobservable subspace of (A, C) is the largest A invariant subspace that is contained in the null-space of C.

3.1.8 Unconstructive Subspaces

A state $x_1 \in \mathfrak{R}^n$ of the system (3.2) and (3.3) is said to be unconstructive if

$$Ce^{At}x_1 = 0$$

For all $t > 0$. That is, a state x_1 is unconstructive if the zero-input response of the system (3.2, 3.3) is zero for all time $t > 0$ whenever $x(0) = x_1$.

Definition 3.6 The constructibility Gramian of the system (3.2) and (3.3) is the matrix

$$W_{cn}(t_o, t_1) = \int_{t_0}^{t_1} e^{AT(\tau - t_0)} C^T C e^{A(\tau - t_0)} d\tau \in \mathfrak{R}^{nxn}, \quad t_o \leq t_1 \tag{3.25}$$

Clearly, we have that $W_0(t_o, t_1) \geq$ for all $t_o, t_1 \in \mathfrak{R}^n$ with $t_o \leq t_1$. The observability Gramian and the constructibility Gramian are related via the identity

$$W_0(t_o, t_1) = e^{AT(\tau - t_0)} W_{cn}(t_o, t_1) e^{AT(\tau - t_0)}.$$

Lemma 3.2 A state $x_1 \in \mathbb{R}^n$ is unconstructive if and only if $x_1 \in N(W_{cn}(0, T))$ for all $T > 0$. By Lemma 2 we see that the unconstructive subspace is given by

$$R_{cn} = \{x_1 \in \mathfrak{R}^n : x_1 \in N(W_{cn}(o, T)) \text{ for all } T > 0\}$$

3.2 Observer Designs for Linear Structures

A state observer is a device that estimates the unknown states of a dynamical system. It utilizes the system model and measurements of the system inputs and outputs for the estimation process. The system model may be represented either by a differential equation (continuous time systems) or by a difference equation (discrete time systems).

Three main quantitative state observers are the Luenberger observer [15], adaptive observer [16], and Kalman filter [17]. Two main techniques are available for observer design. The first one is used for the full-order observer design and produces an observer that has the same dimensions as the original system. The second technique exploits the knowledge of some state space variables available through the output algebraic equation (system measurements) so that a reduced-order observer is constructed only for estimating state space variables that are not directly obtainable from the system measurements.

3.2.1 Luenberger Observer

In the deterministic case, when no random noise is present, the Luenberger observer and its extensions are used for time-invariant systems with known parameters. The theory of observers started with the work of Luenberger [17, 18] so that observers are very often called Luenberger observers. According to Luenberger, any system driven by the output of the given system can serve as an observer for that system.

The equation for the Luenberger observer, in addition to the system dynamics, contains a term that corrects the current state estimates by an amount proportional to the prediction error: the estimation of the current output minus the actual measurement. The inclusion of this correction ensures the stability and convergence of the observer even when the system being observed is unstable

$$\dot{\hat{x}} = A\hat{x} + Bu + L(y - \hat{y}) \tag{3.26}$$

$$\hat{x}(0) = \hat{x}_0$$
$$\hat{y} = C\hat{x} \tag{3.27}$$

where $\hat{x}_0$ is arbitrarily chosen. The error in estimation is defined as

$$e = x - \hat{x}(t) \tag{3.28}$$

The is given by

$$\dot{e} = (A - LC)e \tag{3.29}$$

$$e(0) = x_0 - \hat{x}_0 \tag{3.30}$$

If all the eigenvalues of a matrix $(A - LC)$ are chosen to lie in the left half $s - plane$, then regardless of $e(0)$, $e \to 0$ as $t \to \infty$ exponentially and accurate state estimates are obtained or asymptotically stable. Similarly, the concept for state observers can be extended to nonlinear deterministic continuous time systems. But the design process is far more complicated than the linear systems.

Note that the observer has the same structure as the system plus the driving feedback term that contains information about the observation error. The role of the feedback term is to reduce the observation error to zero (at steady state).

Remark The observer is usually implemented online as a dynamic system driven by the same input as the original system and the measurements coming from the original systems, that is (note $\hat{y} = C\hat{x}$).

Several algorithms have been proposed to realize a stable nonlinear observer. The earliest nonlinear observer designs by Krener and Isidori [18, 19] were based on a set of conditions to linearize the observation error. One of the most complete nonlinear observer designs was proposed by Gauthier et al. [20]. In this observer design, the concept of global nonlinear coordinate changes is used to realize the observer and it guarantees global convergence for uniformly observable inputs.

3.2.2 Kalman Filter

Consider a dynamical system whose state is described by a linear, vector differential equation. The process model and measurement model are represented as

$$\text{Process model: } x = Ax + Bu \tag{3.31}$$

$$\text{Measurement model: } y = Cx + v \tag{3.32}$$

where the vectors u and v are both white noise sequences with zero means and mutually independent:

$$E[u(t)u^T(\tau)] = Q\delta(t - \tau), \tag{3.33}$$

$$E[B(t)u(t)(B(t)u(\tau))^T(\tau)] = BQB^T\delta(t - \tau), \tag{3.34}$$

$$E[v(t)v^T(\tau)] = R\delta(t - \tau), \tag{3.35}$$

$$E[u(t)v^T(\tau)] = 0 \tag{3.36}$$

where
$\delta(t - \tau)$ is the Dirac delta function
$E[\bullet]$ represents expectation value
Superscript T denotes the matrix transpose.
The state estimate equation of the continuous Kalman filter equations is represented as

$$\hat{x} = Ax + K(y - C\hat{x}) \tag{3.37}$$

The propagation of the error for a continuous Kalman filter can be described by the Riccati equation:

$$\dot{P} = AP + PA^T - PC^TR^{-1}CP + BQB^T, \tag{3.38}$$

and the continuous filter gain is obtained through the calculation

$$K = PH^TR^{-1} \tag{3.39}$$

The discrete filter gain and continuous filter gain are related by

$$K = \frac{K_k}{\Delta t} \tag{3.40}$$

where $\Delta t = t_{k+1} - t_k$ represents the sampling period.
When the system reaches steady-state, $\dot{P} = 0$, Eq. (3.38) becomes an algebraic Riccati equation (ARE), which can be solved for the steady-state minimum covariance matrix.

3.2.3 Wiener Filter

The Wiener filter solves the signal estimation problem for stationary signals. The filter was introduced by Norbert Wiener in the 1930s. A major contribution was the use of a statistical model for the estimated signal. The filter is optimal in the sense of the minimum mean square error (MMSE). As we shall see, the Kalman filter solves the corresponding filtering problem in greater generality, for non-stationary signals. We shall focus here on the discrete-time version of the Wiener filter

- x_k the signal to be estimated
- y_k the observed process

which are jointly wide-sense stationary, with known covariance functions: $Rx(k)$ $Ry(k)$ $Rxy(k)$

A particular case is that of a signal corrupted by additive noise:

$$y_k = x_k + n_k \tag{3.41}$$

with x_k, n_k jointly stationary, and $Rx(k)$ $Ry(k)$ $Rxy(k)$ given

Goal: Estimate x_k as a function of y. Specifically:

Find the linear MMSE estimate of x_k based on (all or part of) y_k.

There are three versions of this problem:

a. The non-causal filter:

$$\hat{x}_k = \sum_{m=-\infty}^{\infty} h_{k-m} y_m \tag{3.42}$$

b. The FIR filter:

$$\hat{x}_k = \sum_{m=k-N}^{k} h_{k-m} y_m \tag{3.43}$$

c. The causal filter:

$$\hat{x}_k = \sum_{m=-\infty}^{k} h_{k-m} y_m \tag{3.44}$$

The first problem is the hardest.

Note: For simplicity, in this chapter we consider the scalar case

Consider the FIR filter of length $N + 1$:

$$\hat{x}_k = \sum_{m=k-N}^{k} h_{k-m} y_m = \sum_{i=0}^{N} h_i y_{k-i} \tag{3.45}$$

We need to find the coefficients (h_i) that minimize the MSE:

$$E(x_k - \hat{x}_k)^2 \to min \tag{3.46}$$

To find h_i, we can differentiate the error. More conveniently, start with the orthogonality principle:

$$E[(x_k - \hat{x}_k) y_{k-j}] = 0, \quad j = 0, 1, \dots, N \tag{3.47}$$

this gives

$$\sum_{i=0}^{N} h_i E[y_{k-i} y_{k-j}] = E(s_k y_{k-j}) \tag{3.48}$$

$$\sum_{i=0}^{N} h_i R_y(i-j) = R_{xy}(j) \tag{3.49}$$

$$R_y h = r_{xy} \Rightarrow h = R_y^{-1} r_{xy} \tag{3.50}$$

these are the Yule–Walker equations [21].

We note that $R_y \geq 0$ (positive semi-definite), and non-singular except for degenerate cases. Further, it is a Toeplitz matrix (constant along the diagonals). There exist efficient algorithms (Levinson–Durbin and others) that utilize this structure to efficiently compute h.

3.3 Observer Designs for Nonlinear Structures

3.3.1 Extended Luenberger Observer

The existence of a global nonlinear coordinate change implies uniform observability of the system [22, 23]. Ciccarella et al. [21] proposed several improvements to the observer design by Gauthier et al. [20] to achieve the global asymptotic stability of the observer system. The observer design based on the work by Ciccarella et al. [21] is as follows

For nonlinear systems of the type

$$\dot{x} = f(x) + g(x, u) \tag{3.51}$$

$$y = hx \tag{3.52}$$

where

$$x \in \mathfrak{R}^p$$

$u \in \mathfrak{R}^p$ in the system input

$y \in \mathfrak{R}^m$ are the systems output and the vector functions f, g, h are $\mathbf{C}^\infty$

$$\dot{\hat{x}} = f(\hat{x}) + g(\hat{x}, u) + Q^{-1}(\hat{x}) K(y - \hat{y}) \tag{3.53}$$

$$\hat{y} = h(\hat{x}) \tag{3.54}$$

$Q(\hat{x})$ is the observability matrix defined as

$$Q(\hat{x}) = \frac{d\varphi(x)}{dx} \tag{3.55}$$

$$\varphi(x) = \begin{bmatrix} h(x) \\ L_f h(x) \\ \vdots \\ L^{n-1}{}_f h(x) \end{bmatrix} \tag{3.56}$$

$\varphi(x)$ defines the global change of coordinates. $L_f h(x)$ denotes the Lie derivative of the function h along f and $L^{n-1}{}_f h(x)$ denotes the kth order repeated Lie derivative of the function h along f.

The theory of state observers can similarly be extended to discrete linear and nonlinear systems. The first step in discrete observer realization is the discretization of the system dynamics followed by the observer design in the discrete domain.

3.3.2 Extended Kalman Filter

A straightforward approximation to optimal nonlinear state estimation is to linearize the nonlinear model about a given operating point and apply optimal linear state estimation to the linearized system. The extended Kalman filter computes a state estimate at each sampling time by the use of Kalman filtering on a linearized model of the nonlinear system (bioprocesses). This technique is justified if there exists a sufficiently large neighborhood in which the linearized model is a good representation of the nonlinear system. If, in addition, the disturbances are well represented by zero-mean Gaussian state and measurement noise, the optimal estimate for the linearized system should be a reasonable approximation to the optimal estimate for the nonlinear system [24].

The most common approach is the first-order filter in which the nonlinear system is linearized about the current state estimate at each sampling time using the first-order terms. More complex approaches that attempt to compensate for the inaccuracy caused by linearization, such as iterative and higher-order filters, have also been implemented. These extended Kalman filter techniques are presented in the following sections. Further discussion of extended Kalman filtering is contained in Athans [25], Jazwinski [26], and Gelb [27].

3.3.2.1 First-Order Extended Kalman Filter

For the continuous-time nonlinear system with discrete output measurements in Eq. (3.1), a linearized approximation to the nonlinear model can be obtained by truncating Eq. (3.4) after the first-order terms. This approximate model, linearized about the current state, is used to construct a time-varying Kalman filter at each sampling time. In order to ensure that the partial derivatives exist, the system and measurement functions are restricted to continuous functions in x such that $f, g \in C^1$.

The filtered state estimate is determined from the current output measurement in the same manner as the linear recursive filter in Eq. (3.26)

$$\hat{x}_{k|k} = \hat{x}_{k|k-1} + L_k(y_k - g(\hat{x}_{k|k-1}, k\Delta t))$$
$$\hat{x}_{0|0} = \bar{x}_0 \tag{3.57}$$

The Kalman filter gain, L_k, is computed using the discrete, time-varying Kalman filter formulation in Eq. (3.39) in which the linear system matrix $\bar{C}_k$ is replaced by the linearized measurement function G_k

$$L_k = \hat{P}_{k|k-1} G_K^T (G_k \hat{P}_{k|k-1} G_K^T + R)^{-1}$$
$$G_k = \left. \frac{\partial g(x, t)}{\partial x} \right|_{x = \hat{x}_{k|k-1}, t = k\Delta t} \tag{3.58}$$

The estimated covariance of the state is updated at each sample time due to the contribution of the discrete measurement as follows

$$\hat{P}_{k|k} = (I - L_K G_k)\hat{P}_{k|k-1}$$
$$\hat{P}_{0|0} = Q_0 \tag{3.59}$$

Between sampling times, the state estimate is propagated using the nonlinear system model

$$\dot{\hat{x}}(\tau|k) = f(\hat{x}(\tau|k), u(\tau), \tau), \qquad \hat{x}(k\Delta t|k) = \hat{x}_{k|k}$$
$$\hat{x}_{k+1|k} = \hat{x}((k+1)\Delta t|k) \tag{3.60}$$

The covariance of the state estimate in Eq. (3.60) is propagated between sampling times using the linear differential equation in Eq. (3.38) in which the linear system matrix $A(t)$ is replaced by the linearized system function $F(\tau|k)$. Since there are no output measurements available between sampling times, the contribution to the covariance from the output measurement is removed

$$\dot{\hat{P}}(\tau|k) = F(\tau|k)\hat{P}(\tau|k) + \hat{P}(\tau|k)F(\tau|k)^T, \quad \hat{P}(k\Delta t|k) = \hat{P}_{k|k}$$
$$\hat{P}_{k+1|k} = \hat{P}((k+1)\Delta t|k) + Q$$
$$F(\tau|k) = \left.\frac{\partial f(x, u, t)}{\partial x}\right|_{x=\hat{x}(\tau|k), u=u(k\Delta t), t=r} \tag{3.61}$$

This continuous-time linear approximation of the estimated state covariance is used in Eq. (3.61) to compute the Kalman filter gain. A discrete linear approximation to the covariance can also be used in the Kalman filter gain calculation. It is obtained from the time-varying filtering Riccati equation with the contribution from the output measurement removed

$$\hat{P}_{k|k} = \Phi((k+1)\Delta t, k\Delta t)\hat{P}_{k|k} = \Phi^T((k+1)\Delta t, k\Delta t) + Q \tag{3.62}$$

$$\frac{d\Phi(t, k, \Delta t)}{dt} = F(t|k) \ \Phi(t, k\Delta t), \ \Phi(k\Delta t, k\Delta t) = I \tag{3.63}$$

In this expression, $\Phi((k+1)\Delta t, k\Delta t)$ is the transition matrix of the time-varying linear system obtained by linearization of the nonlinear system. This discrete covariance approximation does not require the solution of a differential equation to determine the covariance, but the state transition matrix must be computed.

The extended Kalman filter for the discrete nonlinear system is similar to that presented for continuous-time systems with discrete measurements.

$$\hat{x}_{k|k} = \hat{x}_{k|k-1} + L_k(y_k - \bar{g}(\hat{x}_{k|k-1}, k)) \tag{3.64}$$

$$L_k = \hat{P}_{k|k-1}\overline{G}_k^T(\overline{G}_k\hat{P}_{k|k-1}\overline{G}_k^T + R)^{-1} \tag{3.65}$$

$$\hat{P}_{k|k} = (I - L_k\overline{G}_k)\hat{P}_{k|k-1} \tag{3.66}$$

$$\hat{x}_{k+1|k} = \bar{f}(\hat{x}_{k|k}, u_k, k) \tag{3.67}$$

$$\hat{P}_{k+1|k} = \overline{F}_k\hat{P}_{k|k-1}\overline{F}_k^T + Q \tag{3.68}$$

The discrete linear system matrices are formed from the partial derivatives of the discrete nonlinear measurement and system functions.

$$\overline{G}_k = \left. \frac{\partial \overline{g}(x, k)}{\partial x} \right|_{x=\widehat{x}_{k|k-1}} \tag{3.69}$$

$$\overline{F}_k = \left. \frac{\partial \overline{f}(x, k)}{\partial x} \right|_{x=\widehat{x}_{k|k},u=u_k} \tag{3.70}$$

The functions $\overline{f}$ and $\overline{g}$ are also restricted to continuous functions in x so that the partial derivatives in Eqs. (3.69) and (3.70) exist. A first-order extended Kalman filter was implemented on an isothermal batch operation in an experimental pilot-scale reactor, a standard nonlinear Kalman filter is used to estimate on-line the concentrations from reactor temperature as the only measured output by DeValliere and Bonvin [28]. Luenberger observers and Kalman–Bucy filters have been applied to estimate non-measurable states in a styrene polymerization reactor by Schuler and Suzhen [29]. Shimizu [30] discusses an algorithm showing the practical procedure for the estimation method utilizing the macroscopic balance and the extended Kalman filter.

Because of the approximations made in the propagation of the covariance matrix in extended Kalman filtering, this covariance matrix can become a poor estimate in some applications. In these cases, updating the filter gain at each sampling period may result in a computational effort that does not improve the performance of the filter [24].

The nominal state trajectory is composed of a single operating point. The robustness of the Kalman filter to model mismatch is sufficient to obtain acceptable performance with this technique in some applications [24].

3.3.3 Asymptotic Observers

For systems that are uniformly observable for any $u(t)$ (i.e. the states of the system can be determined from the output of the system and its derivates, independently of the input) [31], a high-gain observer has been suggested in [32]. One of the advantages of this observer is its excellent robustness properties [32]. By choosing an observer gain k large enough (therefore the name "high-gain") the observer error can be made arbitrarily small. The difficulty in practical applications is, however, the determination of an appropriate value for the observer gain. For too low values, the desired bounds on the observer error cannot be achieved. For values unnecessarily high, the sensitivity to noise increases, thus limiting the practical use [33].

We propose an adapted scheme for the observer gain of the high-gain observer in [9] such that its advantages are retained and the observer output error becomes smaller than the desired target value.

3.3.3.1 High-Gain Observer

The theory of high-gain observers as in [9] assumes that the system is given in observability normal form [34], also called the generalized controller canonical form [35].

In principle, every uniformly observable SISO-system with input u and output y can be transformed into this normal form:

$$\dot{x}_1 = x_2$$
$$\dot{x}_2 = x_3$$
$$\vdots$$
$$\dot{x}_{n-1} = x_n$$
$$\dot{x}_n = \phi(x, u)$$
$$y = x_1 \tag{3.71}$$

with $x = [x_1, \ldots, x_n]^T$ and $u = [u, \dot{u}, u^{(2)}, \ldots, u^{(n)}]^T$.

A possible way to observe such systems (3.71) is the high-gain observer [32]. The structure of the high-gain observer is a simple chain of integrators, each "corrected" by the injection of the output error $(y - \hat{y})$ multiplied by a factor depending on the constant observer gain k:

$$\dot{\hat{x}}_1 = \hat{x}_2 + p_1 k(y - \hat{y})$$
$$\dot{\hat{x}}_2 = \hat{x}_3 + p_2 k^2(y - \hat{y})$$
$$\vdots$$
$$\dot{\hat{x}}_{n-1} = \hat{x}_n + p_{n-1} k^{n-1}(y - \hat{y})$$
$$\dot{\hat{x}}_n = +p_n k^n(y - \hat{y})$$
$$\hat{y} = \hat{x}_1. \tag{3.72}$$

where $\hat{x} = [\hat{x}_1, \ldots, \hat{x}_n]^T$ denotes the estimate of the system's output.

In contrast to the classical Luenberger observer [6], the high-gain observer does not consist of a replica of the system (3.71) plus correction terms as the nonlinearity $\phi(x, u)$ is not modeled. The observer error will be denoted by e with $e(t) = [e_1(t), \ldots, e_n(t)]^T$ with $e_i(t) = x_i(t) - \hat{x}_i(t)$.

The following theorem is proven in [32].

Theorem 3.2   High-Gain Observer
Assume that the system (3.71) exhibits no finite escape time and the nonlinearity ϕ in (1) is bounded, i.e. $\|\phi(x, u)\| \leq \mu$ $\forall x, u$. [31].

If the coefficients $\{p_1, \ldots, p_n\}$ are such that $s^n + \sum_{j=1}^{n} p_j s^{n-j}$ is a Hurwitz polynomial with distinct roots, then for all $d > 0$ and all $\bar{t} > 0$ there exists a finite observer gain $\bar{k}$ such that for all constant $k \geq \bar{k}$ the observed error satisfies:

$$\|e(t)\| \leq d \quad \forall t \geq \bar{t}.$$

This means that by an appropriate choice of the observer gain k the observer error e can be made arbitrarily small in an arbitrarily short time [36].

3.3.4 Adaptive-Gain Observers

3.3.4.1 Adaptive High-Gain Observer

To overcome the difficulty of having to choose the observer gain k, we propose to use a simple adaptation law to find the appropriate observer gain for all $t \geq t_0$:

$$\frac{d}{dt} s(t) = \begin{cases} \varphi \cdot |y(t) - \hat{y}(t)|^2 & \text{for } |y(t) - \hat{y}(t) > \eta| \\ 0 & \text{for } |y(t) - \hat{y}(t) > \eta| \end{cases}$$

$$t_i : s(t_i) = S_i \qquad i = 0, 1, 2, \ldots$$

$$k(t) = S_i \qquad \forall t \in |t_i, t_{i+1} \tag{3.73}$$

where $\eta > 0$, $\varphi > 0$, $v > 0$, S_0 are given and

$$S_{i+1} - S_i = v\, e^{i^2} \quad \forall i \geq 0. \tag{3.74}$$

The idea behind this adaptation is that observer gain k is piecewise constant and takes values S_i [31].

The switch to the new S_i depends on the monotonically increasing parameter s. Whenever s reaches a new threshold S_i the observer gain takes this value S_i. The S_i are predefined as a monotonically increasing sequence such that their growth rate is larger than e^i. One possibility to guarantee this is given in (3.74). For any other choice with growth rate larger than e^i, like

$$S_{i+1} - S_i = v e^{i \ln(i)} \tag{3.75}$$

Thus $k(t)$ is increased step-wise as long as $y - \hat{y}$ lies outside λ and cannot decrease.

We now prove that the high-gain observer (3.72) with a time-varying observer gain k as in adaptation law (3.73) guarantees convergence of the adaptation law and boundedness of the observer error [36].

Theorem 3.3 Assume that system (1) satisfies assumptions (A1) and (A2). If the coefficients $\{p_1, \ldots, p_n\}$ are such that $s^n + \sum_{j=1}^{n} p_j s^{n-j}$ is a Hurwitz polynomial then a high-gain observer (3.72) with the adaptation law (3.73), (3.74) for the observer parameter k achieves for any $\lambda > 0$, $\gamma > 0$, $\beta > 0$ and any S_0:

a) $k(t) < k \infty < \infty \quad \forall t$
b) the total length of time for which the observer output error is larger than λ is finite, i.e. $\exists t \max < \infty : \int_T dt < t_{\max}$, where $T = \{t \,|\, \|y(t) - \hat{y}(t)\| \geq \lambda\}$.

Summarizing the theorems, the adaption of the proposed high-gain observer converges; for a large time the observer output error is smaller than some user-specified bound and the observer error is bounded.

Remark The idea of the proof of the theorems can be used to generalize the results in [35] to the case of no distinct roots.

The proposed adaptive high-gain observer is easy to implement (as only the state-space dimension of the system has to be known) and retains the advantages of the non-adaptive high-gain observer. Robustness is improved by the adaptation law as it enables the user to start with a small observer gain that is increased only as needed. In a non-adaptive scheme, the observer gain is usually chosen in a conservative way, which causes the high-gain observer to perform less well in the presence of output measurement noise than the adaptive high-gain observer [36, 37].

3.3.5 Sliding-Mode Observers

Sliding mode techniques were perhaps originally best known for their potential as a robust control method and evolved from pioneering work in the 1950s in the former Soviet Union. Such a sliding mode control is essentially a particular type of variable structure control system (VSCS) and is characterized by a suite of feedback control laws and a decision rule [37–42]. The decision rule, termed the switching function, has as its input some measure of the current system behavior and produces as an output the particular feedback controller which should be used at that instant in time. In sliding mode control, VSCS is designed to drive and then constrain the system state to lie within a neighborhood of the switching function. There are a number of advantages to this approach [37, 39]. Firstly, the dynamic behavior of the system may be tailored by the particular choice of the switching function. Secondly, the closed-loop response becomes totally insensitive to a particular class of uncertainty in the system. A disadvantage of the method in the domain of control applications has been the necessity to implement a fundamentally discontinuous control signal which, in theoretical terms, must switch with infinite frequency to provide total rejection of uncertainty. Control implementation via approximate, smooth strategies is widely reported, but in such cases, total invariance is routinely lost [36, 39].

In contrast, the application of sliding mode methods to the observer problem is much less mature and has some fundamentally different advantages and disadvantages. The earliest readily available contributions in the area of sliding mode observers appear in the mid-1980s. Walcott et al. [43, 44] define an observer strategy in which the output errors are fed back in both a linear and a discontinuous manner for nonlinear systems in companion form with the objective of ensuring that the so-called "sliding patch," which defines the region in which it is possible for the dynamical observer system to exhibit sliding behavior, is maximized.

A host of application-specific contributions ensued; see for example the work of Sangwongwanich et al. [38], Chen et al. [39], Bartolini et al. [40] as well as work in the area of fault detection and isolation [45].

3.3.5.1 Sliding Mode Observers for Linear Uncertain Systems

Consider first the uncertain dynamical system

$$\dot{x}(t) = Ax(t) + Bu(t) + D\xi(t, y, u)$$
$$y(t) = C_x(t) \tag{3.76}$$

where $A \in \mathfrak{R}^{nxn}$, $B \in \mathfrak{R}^{nxm}$, $C \in \mathfrak{R}^{pxn}$, $D \in \mathfrak{R}^{nxq}$ with $p \geq q$. Assume that the matrices B, C, and D are full rank and the function $\xi : \mathfrak{R} + x\mathfrak{R}^n x\mathfrak{R}^m \to \mathfrak{R}^q$ is unknown but bounded. Let (A, D, C) represent the linear part of the uncertain system given in (3.76) which represents the propagation of the uncertainty through to the output. Define an observer for the uncertain system (3.76) of the form

$$\dot{z}(t) = Az(t) + Bu(t) - G_i Ce(t) + G_n v \tag{3.77}$$

where $e = z x$, v is discontinuous about the hyperplane

$$S_0 = \{e \in \Re^n : Ce = 0\} \tag{3.78}$$

and $G_l, G_n \in \Re^{n \times p}$ are gain matrices whose precise structure is to be determined. A sliding mode observer of the form (3.77) which rejects the uncertainty class in (3.76) exists if and only if the nominal linear system defined by the matrices (A, D, C) satisfies

- rank$(C, D) = q$
- any invariant zeros of (A, D, C)

For a square system, where $p = q$ it should be noted that the above two conditions fundamentally require the triple (A, D, C) to be relative degree one and minimum phase. Note also that these system theoretic conditions depend upon a specific selection of uncertainty channel and the observer design will be directly determined by the uncertainty distribution matrix. When the sliding mode observer is used as a state estimator for a corresponding sliding mode controller, many authors impose the constraint $D = B$, to reflect the fact that the sliding mode controller has the ability to reject all uncertainty implicit in the channels of the system input [45]. In this case, the existing conditions are implicit in the properties of the original system triple (A, D, C). The following canonical form as presented in Edwards et al. [45] is useful to explore the pertinent characteristics of sliding mode observers for linear, uncertain systems.

A change of coordinates exists so that the triple with respect to the new coordinates (A, D, C) has the following structure:
The system triple $(\tilde{A}, \tilde{D}, \tilde{C})$ is now in the form

$$\dot{x}_1(t) = A_{11}x_1(t) + A_{12}y(t) + B_1 u(t)$$
$$\dot{y}(t) - A_{21}x_1(t) + A_{22}y(t) + B_2 u(t) + D_2\xi \tag{3.79}$$

where $x_1 \subset \Re^{(n-p)}$, $y \subset \Re^p$ and the matrix A_{11} is stable, hhere $A_{11} \in \Re^{(n-p) \times (n-p)}$ and $A_{211} \in \Re^{(n-p) \times (n-p)}$. When partitioned these matrices have the structure

$$A_{11} = \begin{bmatrix} A_{11}^0 & A_{12}^0 \\ 0 & A_{22}^0 \end{bmatrix} \quad \text{and} \quad A_{211} = [0 \ A_{21}^0] \tag{3.80}$$

where $A_{11}^0 \in \Re^{r \times r}$ and $A_{21}^0 \in \Re^{(pq) \times (n \ p \ r)}$ for some $r > 0$ and the pair A_{22}^0, A_{21}^0 is completely observable.

Define the corresponding observer by

$$\hat{x}_1(t) = A_{11}\hat{x}_1(t) + A_{12}\hat{y}(t) + B_1 u(t) - A_{12}e_y(l)$$
$$\dot{y}(t) - A_{21}\hat{x}_1(t) + A_{22}\hat{y}(t) + B_2 u(t) - (A_{22} - A_{22}^s)e_y(t) + v \tag{3.81}$$

where A_{22}^s is a stable design matrix and $c_y = \hat{y} - y$. Let $P_2 \in \mathfrak{R}^{p \times p}$ be the symmetric positive definite Lyapunov matrix for A_{22}^s then the discontinuous vector v is defined by

$$v - \begin{cases} -p(t, y, u)\|D_2\|\frac{P_2 e_y}{\|P_2 e_y\|} & if \ e_y \neq 0 \\ 0 & otherwise \end{cases} \tag{3.82}$$

where the scalar function $p : \mathfrak{R}_+ x \mathfrak{R}^p x \mathfrak{R}^m \to \mathfrak{R}_+$ satisfies

$$p(t, y, u) \geq r_1 \|u\| + \alpha(t, y) + \gamma_0 \tag{3.83}$$

and γ_0 is a positive scalar. If the state estimation error $e_1 = \hat{x}_1 - x_1$, then it is straightforward to show

$$\dot{e}_1(t) - A_{11} e_1(t) \tag{3.84}$$

$$\dot{e}_y(t) - A_{21} e_1(t) + A_{22}^s e_y(t) + v - D_2 \xi \tag{3.85}$$

Define $Q_1 \in \mathfrak{R}^{(n-p) \times (u-p)}$ and $Q_2 \in \mathfrak{R}^{p \times p}$ as symmetric positive definite design matrices and define $P_2 \in \mathfrak{R}^{p \times p}$ as the unique symmetric positive definite solution to the Lyapunov equation

$$P_2 A_{22}^s + (A_{22}^s)^T P_2 = -Q_2 \tag{3.86}$$

Using the computed value of P_2 define

$$\dot{Q} - A_{21}^T P_2 Q_2^{-1} P_2 A_{21} + Q_1 \tag{3.87}$$

Note that $\hat{Q} = \hat{Q}^T > 0$ and let $P_1 \in \mathfrak{R}^{(n-p) \times (n-p)}$ be the unique symmetric positive definite solution to the Lyapunov equation

$$P_1 A_{11} + A_{11}^T P_1 = -\hat{Q} \tag{3.88}$$

Taking the quadratic form

$$V(e_1, e_y) = e_1^T P_1 e_1 + e_y^T P_2 e_y \tag{3.89}$$

as a candidate Lyapunov function it can be shown that the error system is quadratically stable.

Further, consideration of the quadratic form

$$V_s(e_y) = e_y^T P_2 e_y \tag{3.90}$$

shows that an ideal sliding motion takes place on (3.78) and the output error e_y enters $\Omega = \{(e_1, e_y) \ : \ \|A_{21} e_1\| < \|D_2\| \gamma - \eta\}$ where η is a small positive scalar in finite time and remains there. An ideal sliding motion takes place on (3.78) in finite time. This finite time convergence property is a key advantage of sliding mode observer schemes. Many other observer design paradigms guarantee only asymptotic properties.

If $\hat{x}$ represents the state estimate for x and $e = \hat{x} - x$ a robust observer can conveniently be written as

$$\dot{x}(t) = A\hat{x}(t) + Bu(t) - G_l C e(t) + G_n v \tag{3.91}$$

Observability of the pair (A, C) was assumed in some of the early papers published in the area, but this has later been disproved. This is a further key distinguishing characteristic of a sliding mode observer relative to other observer design approaches which often routinely impose observability restrictions on the system representation [37].

3.3.5.2 Nonlinear Approaches to Sliding Mode Observer Design

Early contributions in this area were developed independently by Walcott et al. [43, 44], Slotine et al. [32, 36], where the latter team considered a more extended class of systems. As in Slotine et al. [36], consider a nonlinear system in companion form

$$x^{(n)} - f(x, t) \tag{3.92}$$

where $f(x, t)$ is a nonlinear, uncertain function of the system state and x_1 is the single measurement available. Define a corresponding sliding mode observer

$$\hat{\dot{x}}_1 = -a_1 e_1 + \hat{x}_2 - k_1 \ \mathrm{sgn}(e_1)$$
$$\hat{\dot{x}}_2 = -a_2 e_2 + \hat{x}_3 - k_2 \ \mathrm{sgn}(e_1)$$
$$\dots$$
$$\hat{\dot{x}}_n = -a_n e_1 + \hat{f} - k_n \ \mathrm{sgn}(e_1) \tag{3.93}$$

where $e = \hat{x}_1 - x_1, \hat{f}$ is an estimate of $f(x, t)$ and the constants a_1 are chosen as for a classical Luenberger observer to ensure asymptotic error decay of a corresponding linearized system representation, where $k_i = 0$. The corresponding error dynamics are given by

$$\dot{e}_1 = -a_1 e_1 + e_2 - k_1 \ \mathrm{sgn}(e_1)$$
$$\dot{e}_2 = -a_2 e_2 + e_3 - k_2 \ \mathrm{sgn}(e_1)$$
$$\dots$$
$$\dot{e}_n = -a_n e_1 + \Delta f - k_n \ \mathrm{sgn}(e_1) \tag{3.94}$$

where $\Delta f = \hat{f} - f$ is assumed bounded and

$$k_n \geq |\Delta f| \tag{3.95}$$

The sliding condition $\frac{d}{dt}(e_1)^2 < 0$ is satisfied in the region

$$e_2 \leq k_1 = -a_1 e_1 \ \text{if} \ e_1 > 0$$
$$e_2 \geq -k_1 - a_1 e_1 \ \text{if} \ e_1 < 0 \tag{3.96}$$

From (3.95), when a sliding mode is attained on $e_1 = 0$ it follows that

$$e_2 - k_1 \ \mathrm{sgn}(e_1) = 0 \tag{3.97}$$

and therefore

$$\dot{e}_2 = e_3 - \frac{k_2}{k_1} e_2$$
$$\dots$$
$$\dot{e}_n = \Delta f - \frac{k_n}{k_1} e_2 \tag{3.98}$$

The values α_1 are thus seen to only affect the dynamic performance prior to reaching this region, which is often called the sliding patch, and the dynamics on the patch are determined by

$$
\left| \lambda I_{n-1} - \begin{pmatrix} -\frac{k_2}{k_1} & 1 & 0 & \cdots & 0 \\ -\frac{k_3}{k_1} & 0 & 1 & \cdots & 0 \\ & & & \cdot & \\ & & & \cdots & 1 \\ -\frac{k_n}{k_1} & 0 & 0 & \cdots & 0 \end{pmatrix} \right|
$$

(3.99)

Assuming k_n is selected as a constant ratio with k_1 and that the poles defining the dynamics on the patch are critically damped i.e. are real and equal to some constant value γ, then Slotine et al. [34] show that

$$
|e_2^{(i)}| \le (2\gamma)^i k_1 \quad i = 0, \ldots, n-2
$$

(3.100)

from which the precision of the state estimates can be determined.

Overall, measurement or estimation of bioprocess states is critical for process control and optimization applications. However, certain disturbances or unknown inputs can generate significant model–plant mismatch if not considered by the process models. The model can be used in terms of substrates, biomass and metabolites concentrations. For example, the estimation is made from the measurement of the volatile fatty acid concentration at the outflow stream, the biomass concentration can be used as the output information because it is feasible to measure it on-line by regular sensors [46–51].

List of Figure

Figure 3.1   A general estimation scheme (*x*: state variables, *π*: parameters, *y*: outputs) [9]. *Source*: Reproduced with permission of Elsevier.   *111*

References

1 Bouraoui, I., Farza, M., Menard, T. et al. (2016). *On-Line Estimation of the Reaction Rates from Sampled Measurements in Bioreactors IFAC DYCOPS-CAB*. Trondheim, Norway: NTNU.

2 Kadlec, P., Gabrys, B., and Strandt, S. (2009). Data-driven soft sensors in the process industry. *Computers & Chemical Engineering* 33, 4: 795–814.

3 Chachuat, B. and Bernard, O. (2005). Probabilistic observers for a class of uncertain biological processes. *International Journal of Robust and Nonlinear Control*, Wiley 16: 157–171.

4 Bernard, O. and Gouzé, J.-L. (2002). *State Estimation for Bioprocesses* (ed. A. Agrachev), 813–855. Trieste, Italy: ICTP Lecture Notes.

5 López-Pérez, P.A., Neria-González, M.I., and Aguilar-López, R. (2016). Concentrations monitoring via software sensor for bioreactors under model parametric uncertainty:

application to cadmium removal in an anaerobic process. *Alexandria Engineering Journal* 55 (2): 1893–1902.

6 Bastin, G. and van Impe, J. (1995). Nonlinear and adaptive control in biotechnology: a tutorial. *European Journal of Control* 1 (1): 1–37.

7 Deza, F., Bossanne, D., Busvelle, E. et al. (1993). Exponential observers for nonlinear systems. *IEEE Transactions on Automatic Control* 38 (3): 482–484.

8 Asbjørnsen, O.A. (1972). Reaction invariants in the control of continuous chemical reactors. *Chemical Engineering Science* 27: 709–717.

9 Bastin, G. and Dochain, D. (1990). *On-Line Estimation and Adaptive Control of Bioreactors*. Amsterdam, The Netherlands: Elsevier.

10 Moore, B. (1981). Principal component analysis in linear systems: controllability, observability, and model reduction. *Transactions on Automatic Control* 26: 17–32.

11 Ritt, J. (1950). *Differential Algebra*. American Mathematical Society.

12 Kolchin, E. (1973). *Differential Algebra and Algebraic Groups*. New York, NY: Academic Press.

13 Perko, L. (2001). *Differential Equations and Dynamical Systems*, 3e. New York, NY: Springer-Verlag.

14 Luenberger, D.G. (1966). Observers for multivariable systems. *IEEE Transactions on Automatic Control* 11: 190–197.

15 Bernard, O., Sallet, G., and Sciandra, A. (1998). Nonlinear observers for a class of biological systems. Application to validation of a phytoplanktonic growth model. *IEEE Transactions on Automatic Control* 43: 1056–1065.

16 Luenberger, D.G. (1964). Observing the state of a linear system. *IEEE Transactions on Military Electronics* 8: 74–80.

17 Luenberger, D.G. (1971). An introduction to observers. *IEEE Transactions on Automatic Control* 16: 592–602.

18 Kerner, A. and Isidori, A. (1983). Linearization by output injection and nonlinear observers. *Systems and Control Letters* 3 (1): 47–52.

19 Isidori, A. (1995). *Nonlinear Control Systems*. New York, NY: Springer-Verlag.

20 Gauthier, J., Hammouri, H., and Othman, S. (1992). A simple observer for nonlinear systems. *IEEE Transactions on Automatic Control* 37 (6): 875–880.

21 Ciccarella, G., Mora, M.D., and Germani, A. (1993). A Leunberger-like observer for nonlinear systems. *International Journal of Control* 57 (3): 537–556.

22 N. Shimkin. (2009) *Estimation and Identification in Dynamical Systems (048825) Lecture Notes, Fall, Technion*. Israel Institute of Technology, Department of Electrical Engineering.

23 Praveen c. Muralidhar. (2006) *Observer synthesis for linear/nonlinear dynamical systems subject to measurement delays*. The University of Texas at Arlington in Partial Fulfillment of the Requirements for the Degree of the University of Texas at Arlington.

24 Muske, K. and Edgar, T. (1997). *Nonlinear State Estimation, in Nonlinear Process Control* (eds. M. Henson and D. Seborg), 311–370. New Jersey, Chapter 6: Pentice Hall.

25 Athans, M., Wishner, R., and Bertolini, A. (1968). Suboptimal state estimation for continuous-time nonlinear systems from discrete Noisy measurements. *IEEE Transactions on Automatic Control* 13: 504–514.

26 Jazwinski, A.H. (1970). *Stochastic Processes, and Filtering Theory*. New York, NY: Academic Press.

27 Gelb, A. (ed.) (1974). *Applied Optimal Estimation*. The MIT Press.

28 de Vallière, P., Agarwal, M., and Bonvin, D. (1988). Experimental estimation of concentrations from reactor temperature measurement. *IFAC Proceedings* 21 (7): 183–188.

29 Schuler, H. (1980). *Estimation of states in a polymerization reactor proc*. In: *IFAC PRP 4 Automation*, 369–376. Ghent, Belgium: Elsevier.

30 Shimizu, H., Takamatsu, T., Shioya, S., and Suga, K. (1989). An algorithmic approach to constructing the on-line estimation system for the specific growth rate. *Biotechnology and Bioengineering* 33: 354–364.

31 E. Bullinger and F. Allgower. (2010) An adaptive high-gain observer for nonlinear systems. *IEEE Conference on Decision and Control, San Diego, California*, **36**.

32 Slotine, J.J.E., Hedrick, J.K., and Misawa, E.A. (1986). On sliding observers for nonlinear systems. In: *Proceedings of the American Control Conference*, 1794–1800.

33 Itkis, U. (1976). *Control Systems of Variable Structure*. New York, NY: Wiley.

34 Utkin, V.I. (1992). *Sliding Modes in Control Optimisation*. Berlin: Springer-Verlag.

35 Bullinger, E., Ilchmann, A., and Allgöwer, F. (1998). Nonlinear control systems design 1998: a proceedings volume. In: *4th IFAC Symposium, Enschede, the Netherlands*, vol. 2 (ed. H.J.C. Huijbert), 781–786. Oxford: Pergamon Press.

36 Slotine, J.J.E., Hedrick, J.K., and Misawa, E.A. (1987). On sliding observers for nonlinear systems. *ASME Journal of Dynamic Systems and Control* 109: 245–252.

37 Sarah K. Spurgeon. (2015) *Sliding mode observers - historical background and basic introduction*. School of Engineering and Digital Art University of Kent, UK Spring School, Aussois.

38 Sangwongwanich, S., Yonemoto, T., Furuhashi, T., and Okuma, S. (1990) *Design of sliding observer for robust estimation of rotor flux of induction motors*. Proceedings of IPEC, Tokyo, Japan.

39 Chen, D.S., Utkin, V.I., Zarei, S., and Miller, J.M. (2000). Real-time implementation of sliding mode observer for synchronous rectification of the automotive electrical power supply system. *ASME Journal of Dynamic Systems Measurement and Control* 122: 594–598.

40 Bartolini, G., Damiano, A., Gatto, G. et al. (2003). Robust speed and torque estimation in electrical drives by second order sliding modes. *IEEE Transactions on Control Systems Technology* 11: 84–90.

41 Edwards, C., Spurgeon, S.K., and Patton, R.J. (2000). Sliding mode observers for fault detection and isolation. *Automatica* 36: 541–553.

42 Aguilar-López, R., Martínez-Guerra, R., and Maya-Yescas, R. (2003). State estimation for partially unknown nonlinear systems: a class of integral high gain observers. *IEE Proceedings Control Theory and Application* 150: 240–244.

43 Walcott, B.L., and Zak, S.H. (1986) *Observation of dynamical systems in the presence of bounded nonlinearities/uncertainties*. Proceedings of the 25th IEEE CDC, Athens, Greece.

44 Walcott, B.L., Corless, M.J., and Zak, S.H. (1987). Comparative study of state observation techniques. *International Journal of Control* 45: 2109–2132.

45 Edwards, C. and Spurgeon, S.K. (1998). *Sliding Mode Control: Theory and Applications*. London: Taylor and Francis.

46 Robertson, D., Lee, J., and Rawlings, J. (1996). A moving horizon-based approach for least-squares state estimation. *AIChE Journal* 42: 2209–2224.

47 Alzate-Ibanez, A. (2018). States and variables estimation in an upflow anaerobic sludge blanket reactor for the leachate wastewater treatment using nonlinear observers. *Revista Mexicana De Ingeniería Química* 17 (2): 723–738.

48 Martínez, N., Poznyak, A., & Chairez, I. (2015). Software sensor for microalgae culture: Theoretical development, numerical solutions and real time implementations. En Microalgae and Other Phototrophic Bacteria: Culture, Processing, Recovery and New Products

49 Pan, X., Raftery, J.P., Botre, C. et al. (2019). Estimation of unmeasured states in a bioreactor under unknown disturbances. *Industrial and Engineering Chemistry Process Design and Development* 58: 2235–2245.

50 Hong, J., Laflamme, S., Dodson, J. et al. (2018). Introduction to state estimation of high-rate system dynamics. *Sensors* 18 (1): 217.

51 Ekramian, M., Sheikholeslam, F., Hosseinnia, S. et al. (2013). Adaptive state observer for lipschitz nonlinear systems. *Systems & Control Letters* 62: 319–323.

52 M. Agarwal, D. Bonvin. (1989) *Limitation of extended Kalma filter for batch reactors*. IFAC Symposium DYCORD '89, Maastricht 2.

53 Spurgeon, S.K. (2008). Sliding mode observers: a survey. *International Journal of Systems Science* 39 (8): 751–764.

4

Control of Bioprocess

This chapter introduces the reader to the controllability concepts that are the basis of design automatic feedback control schemes for bioprocess. A clear explanation is developed from the classical linear schemes to advanced robust algorithms with application in bioprocess systems. From the environmental point of view, researchers are focusing on design and optimization of control systems, through simulation and modeling of biological processes, devoted to avoiding and/or remedying environmental pollution. Currently some bioprocess plants use very simple control technologies such as simple programmable logic controller (PLC)-techniques, pneumatic control, manual control, and simple linear control. Several tuning procedures for proportional-integral-derivative (PID) algorithms have been proposed in the literature to retain good performance. The conventional PID algorithms are still widely used in the industry since they are relatively easy to implement, and time tested. However, performance may degrade when they are applied to highly nonlinear processes, which are the fact rather than the exception in the chemical and biochemical process industry.

4.1 The Control Idea

Generally, standard PID algorithms with fixed parameters may perform poorly when the process gain varies substantially with operating conditions. Nonlinear control such as nonlinear model predictive control (MPC) should be used as the approach adapts the proportional-integral (PI) settings continuously online to force the resulting closed-loop response to satisfy the predefined performance specification. The approach can improve the control performance even if no a priori knowledge of good values for the PI settings is available. The special features of the proposed automatic tuner can be summarized as follows: it incorporates closed-loop prediction criteria; the performance specification is expressed in the form of time domain constraints, which makes it more appealing for the practitioner. However, the performance of these methodologies may not be satisfactory when applied to nonlinear processes. For biological processes, several kinds of control strategies have been proposed, such as, adaptive, optimal, H∞, linearizing, and neuro-controllers [1–3]. However, they have shown adequate performance for a class of reacting systems, some of them are coupled with optimizing routines or are model based, their main drawbacks are the over-parameterization and lack of robustness under

Control in Bioprocessing: Modeling, Estimation and the Use of Soft Sensors, First Edition.
Pablo Antonio López Pérez, Ricardo Aguilar López, and Ricardo Femat.
© 2020 John Wiley & Sons Ltd. Published 2020 by John Wiley & Sons Ltd.

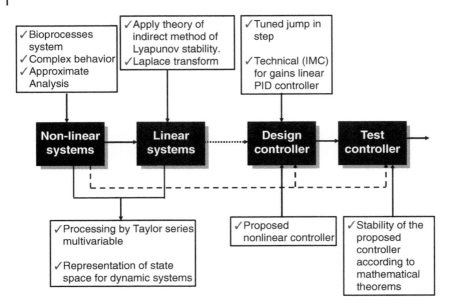

Figure 4.1 The control idea in bioprocess.

model uncertainties. Today, the gain scheduled scheme and the adaptive PI control are two alternatives used to handle nonlinear processes [4]. Nonlinear system control has been widely studied over recent years; the main topics have been the stability of dynamic systems [5, 6], predictive control theory [7] and nonlinear variable gain [8–10].

Generally, PI or PID control strategies can be used for simply managing process variables such as temperature and pressure. pH is somewhat more sophisticated to avoid oscillation and offset, dissolved oxygen (DO) is managed by cascade control, critical variables, such as nutrient feed rate(s) may not have simple hardware sensors available (glucose monitoring) [11].

A general methodology for the design of control laws for bioprocesses is presented in Figure 4.1. This figure relates in general terms to a nonlinear system, the different tools to design, implement, and test the law of nonlinear and linear control based on a mathematical model of reference.

4.1.1 General Definitions

Definition 4.1 Input variables are those that independently stimulate the system and can there by induce change in the internal conditions of the process.

Definition 4.2 Output variables are those by which one obtains information about the internal state.

Definition 4.3 Control action is taken based on information that there is a process upset which occurs when the process output variable deviates from the set-point.

Definition 4.4 Standard feedback controllers

   i. proportional controller (P)
  ii. proportional + integral controller (PI)
 iii. proportional + integral + derivative controller (PID)

Definition 4.5 State, state-variables, state-space, trajectory: state of a system at t_0: includes the minimum information necessary to completely specify the condition of the system at t_0 and allow determination of all system outputs at $t > t_0$ when inputs up to time t are specified.

State: a set of state variables

$$\text{State}: x = \begin{bmatrix} x_i \\ x_j \end{bmatrix} \text{ (state vector)} \tag{4.1}$$

State space:

$$\text{Set of all possible } x = \begin{bmatrix} x_i \\ x_j \end{bmatrix} \tag{4.2}$$

All possible paired values $(x_i(1), x_j(1))$; $(x_i(2), x_j(2))$; ...)
Trajectory of the state

$$x = \begin{bmatrix} x_i(20) \\ x_j(20) \end{bmatrix} \quad x = \begin{bmatrix} x_j(21) \\ x_i(21) \end{bmatrix}$$

a curve in two-dimensional space.

$$\text{In general } x = \begin{bmatrix} x_1 \\ \vdots \\ x_n \end{bmatrix} \tag{4.3}$$

In the theory of continuous linear time-invariant (LTI) dynamical control systems the most popular and the most frequently used mathematical model is given by the following differential state equation and algebraic output Eq. (4.4).

Uniform structure (form) for all linear-systems: despite the order, the numbers of inputs and outputs, and forms of the input functions [12]

$$\left. \begin{aligned} \dot{x} &= Ax + Bu \\ y &= Cx + Du \end{aligned} \right\} \begin{aligned} &\text{state equation} \\ &\text{output equation} \end{aligned} \tag{4.4}$$

Here

   i. $x = (x_1, \ldots, x_n)^T$: state vector
  ii. $u = (u_1, \ldots, u_m)^T$: input (or control) vector
 iii. $y = (y_1, \ldots, y_p)^T$: output vector
 iv. $A \in \mathfrak{R}^n$: is the state matrix
  v. $B \in \mathfrak{R}^{n \times m}$: is the input matrix
 vi. $C \in \mathfrak{R}^{p \times n}$: is the output matrix
 vii. $D \in \mathfrak{R}^{p \times m}$: is the feedthrough (or feedforward) matrix
viii. the systems (4.4) is called a linear time-invariant system (LTI).

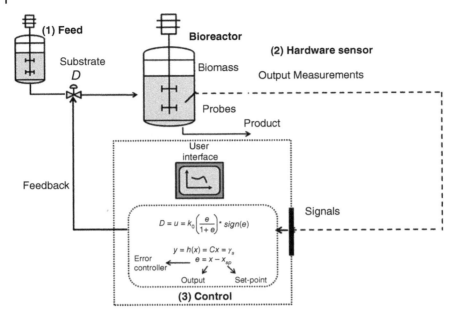

Figure 4.2 Schematic of the nonlinear PI controller.

Controllability is an important property of a control system, and the controllability property plays a crucial role in many control problems, such as the stabilization of unstable
systems control.

Definition 4.6 *Nonlinear Controller Implementation in Bioprocess*
A classic representation of a real-time implementation of a bioprocess control law is shown
in Figure 4.2. This figure consists of three stages that describe the operation of an open-loop
and closed-loop bioreactor using the measurement of a state variable available in the bioreactor to feed back to the controller to drive the variable(s) to maintain an ideal concentration
(set-point) in the bioreactor for maximum performance of the desired product [13].

4.1.2 Controllability of Input/State/Output Systems

Definition 4.7 Dynamical system (4.4) is said to be controllable if for every initial condition $x(0)$ and every vector $x_1 \in \Re^n$, there exist a finite time t_1 and control $u \in \Re^m$, $t \in [0, t]$,
such that $x(t_1) = x_1$.

Controllability can be easily computed by means of the following algebraic criteria:
the system is controllable if and only if the matrix presented in the Eq. (4.5) has full rank

$$C = [B\,AB\,A^2B\ldots A^{n-1}B] \tag{4.5}$$

This matrix is called the controllability matrix.

Theorem 4.1 Dynamical system (4.4) is controllable if and only if $rank C = n$.

Therefore, it is possible to define the so-called output controllability for the output vector
$y \in \Re^p$ of the dynamical system.

Definition 4.8 Dynamical system (4.4) is said to be output controllable if for every $y(0)$ and every vector $y_1 \in \mathfrak{R}^p$, there exist a finite time t_1 and control $u_1 \in \mathfrak{R}^m$, that transfers the output from $y(0)$ to

$$y_1 = y(t_1) \tag{4.6}$$

For a linear continuous-time system, like (4.4), described by matrices A, B, C, and D, the output controllability matrix is defined

$$oC = [CB\ CAB \dots CA^{n-1}BD] \tag{4.7}$$

Theorem 4.2 Dynamical system (4.4) is output controllable if and only if rank $oC = p$

Theorem 4.3 The output controllability character is invariant under feedback.

4.1.3 Steady-Output Controllability

Within the linear system theory, it is often asked whether it is possible that the state-steady outputs converge to a constant value.

Definition 4.9 A vector is called constant steady-state output controllable if there exists an input constant vector u such that

$$\lim_{t \to \infty} y(t) = K \tag{4.8}$$

where K is a $p \times 1$ constant output vector.
   Taking Laplace transforms to the system

$$\dot{x} = Ax + Bu$$
$$y = Cx \tag{4.9}$$

reformulating this definition in the following manner.

Proposition 4.1 A constant output vector K is steady output controllable if there exists an input $u(s) = \frac{k}{s}$.
   Clearly a necessary condition for constant steady output controllability of the system is that the system be stable. The concept of stability is very important in systems theory.
   Remember that a system is stable if and only if

$$rank \begin{bmatrix} S_0 I_n & -AB \\ C & 0 \end{bmatrix} = rank \begin{bmatrix} SI_n & -AB \\ C & 0 \end{bmatrix}, \quad \forall S_0 \in \mathfrak{R}^+ \tag{4.10}$$

Proposition 4.2 A necessary and sufficient condition for constant steady-state output controllability of a stable system is

$$rank \begin{bmatrix} A & B \\ C & 0 \end{bmatrix} = n + min\{m, p\} \tag{4.11}$$

No all systems are stable but sometimes, it is possible to stabilize them by means of a feedback or/and an output injection, concretely, it can be said that a system as (4.9) is stabilizable if and only if there exist a feedback $F \in \mathcal{R}^{mxn}$ or/and output injection $J \in \mathcal{R}^{nxp}$ such that the system closed loop system

$$\dot{x} = (A + BF + JC)x + Bu$$

$$y = Cx \tag{4.12}$$

is stable.

4.1.4 Linear Controllability Analysis LTI Test

4.1.4.1 Controllable and Reachable Subspaces

Definition 4.10 *(Reachable Systems)*
Considered dynamical system (4.4), given two times $t_1 > t_0 \geq 0$, the systems (4.4) or simply the pair (A, B), is (completely state) reachable on $[t_0, t_1]$ if $R[t_0, t_1] = \mathcal{R}^n$, i.e. if the origin can be transferred to every state.

Definition 4.11 *(Controllability Systems)*
Considered dynamical system (4.4), given two times $t_1 > t_0 \geq 0$, the systems (4.4) or simply the pair (A, B), is (completely state) reachable on $[t_0, t_1]$ if $\wp[t_0, t_1] = \mathcal{R}^n$, i.e. if the origin can be transferred to every state [14].
Notes:

i. For continuous-time LTI systems $R[t_0, t_1] = \wp[t_0, t_1]$, and therefore one often talks only about controllability.
ii. A system that is not controllable is called uncontrollable.

4.1.4.2 Controllable Matrix Test

Theorem 4.4 The LTI (4.4) system is controllable if and only if

$$rankC = n \tag{4.13}$$

4.1.4.3 Eigenvector Test for Controllability

Theorem 4.5 The LTI (4.4) system is controllable if and only if there is no eigenvector of A' in the kernel of B'.
Notes:

i. The eigenvalues corresponding to left-eigenvector of A in the left-kernel of B are called the uncontrollable modes and the remaining ones are called the controllable modes [13, 14].

$$C'x = \begin{bmatrix} B' \\ B'A' \\ \vdots \\ B'(A')^{n-1} \end{bmatrix} \quad x = \begin{bmatrix} B'x \\ \lambda B'x \\ \vdots \\ \lambda^{n-1}B'x \end{bmatrix} = 0 \tag{4.14}$$

ii. The LTI (4.4) system is not controllable, then there must exist an eigenvector of A' in the kernel of B'.

4.1.4.4 Popov–Belevitch–Hautus

Theorem 4.6 The LTI (4.4) system is controllable if and only if

$$rank[A - \lambda I] = n \quad \forall \lambda \in \mathfrak{I} \tag{4.15}$$

4.1.4.5 Lyapunov Test for Controllability

Theorem 4.7 The LTI (4.4) system is controllable if and only if there is a unique positive-definite solution W to the following Lyapunov equation

$$AW + WA' = -BB'/AWA' - W = -BB' \tag{4.16}$$

$$W = \int_0^\infty e^{At} BB' e^{A't} d\tau$$

$$= \lim_{t_1 - t_0 \to \infty} W_R(t_0, t_1)/W = \sum_{\tau=0}^\infty A^\tau BB'(A')^\tau = \lim_{t_1 - t_0 \to \infty} W_R(t_0, t_1) \tag{4.17}$$

Since W is positive-definite, the equation can be written as

$$\overline{A}' W + W\overline{A} = -Q\overline{A} := A' \ Q := BB' \tag{4.18}$$

$$W(0, 1) > 0 \tag{4.19}$$

due to controllability.

4.1.5 Stabilizability

Definition 4.12 Considered dynamical system (4.4), the pair (A, B), if it is algebraically equivalent to a system in the standard form for uncontrollable systems which is $n = \overline{n}$ (see Table 4.1).

4.2 Controllers for Linear Systems

- *Single-input single-output* (SISO). The system has one input u and one output x.
- *Multiple-input multiple-output* (MIMO). The system has multiple input u and multiple output x.
- *Single-input multiple-output* (SIMO). Can be regarded as several SISO systems.
- *Multiple-input single-output* (MISO). Can be regarded as several SISO systems.

4.2.1 Linear Feedback

Design a device that makes the process (A, B, C) asymptotically stable by manipulating the input u to the process, see Figure 4.3.

If measurements of the state vector are available, we can set

$$\mu(k) = k_1 x_1(k) + k_2 x_2(k) + \cdots + k_n x_n(k) + \sigma(k) \tag{4.20}$$

$\sigma(k)$ is an exogenous signal exciting the closed-loop system.

Table 4.1 Review and applications of the properties the controllability and observability test for LTI systems.

Review		Test			
Properties	Definition	Rank	Eigenvector	Lyapunov	Applications
Controllability	Every initial state can be taken to the origin in finite time	$rank\underbrace{[B\ AB\ \ldots A^{n-1}B]}_{\text{controllability matrix}} = n$	No eigenvector of A' in the kernel of B'	$\exists W > 0$: $AW + WA' + BB' = 0$	[15–18]
Stabilizability	Asymptotically stable uncontrollable component of the state in standard form for uncontrollable systems	–	No unstable or marginally stable eigenvector of A' in the kernel of B'	$\exists P > 0$: $AP + W\dot{P} + BB' < 0$	[19–21]
Observability	Every non-zero initial state results in a non-zero output	$rank\underbrace{\begin{bmatrix} C \\ CA \\ \vdots \\ CA^{n-1} \end{bmatrix}}_{\text{observability matrix}} = n$	No eigenvector of A in the kernel of C	$\exists W > 0$: $WA + A'W + C'C = 0$	[22–25]

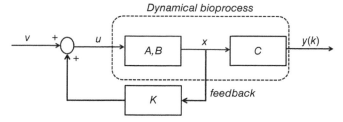

Figure 4.3 Stabilization by state feedback.

It is necessary to find a feedback gain $K = [k_1\ k_2...k_n]$ that makes the closed-loop system asymptotically stable.

4.2.2 Proportional, Proportional-Integral, Proportional-Integral-Derivative

We want to make the system stable and controllable with a controller. The PID controller is a simple controller that may achieve this goal. The PID controller is often analyzed in the frequency domain.

$$\mu = k_p e + k_i \int e(\tau)d\tau + k_d \dot{e} \tag{4.21}$$

i. Proportional
- a pure proportional controller will have a steady-state error
- adding an integration term will remove the bias
- high gain (k_p) will produce a fast system
- high gain may cause oscillations and may make the system unstable
- high gain reduces the steady-state error.

ii. Integral
- removes steady-state error
- increasing k_i accelerates the controller
- high k_i may give oscillations
- increasing k_i will increase the settling time.

iii. Derivative
- larger k_d decreases oscillations
- improves stability for low values of k_d
- may be highly sensitive to noise if one takes the derivative of a noisy error
- high noise leads to instability.

4.2.3 Optimal Control

- Optimal control is another control approach to PID.
- The idea is to specify a cost function and then find the optimal input.
- The dynamics of the system is used to design the controller.

- For nonlinear systems it is not always possible to find the optimal solution.
- A special case is for linear systems with a quadratic cost function.
- The optimal controller must have all states as input.
- Most often used with an observer to estimate the states that are not measured for all future time steps in order to find the optimal solution.

Consider the system (4.22) with $x \in \mathfrak{R}^n$, $u \in \mathfrak{R}^m$, and $x(0) = x_0$:

$$\dot{x} = Ax + Bu \qquad (4.22)$$

Find open loop control $u(\tau)$, $\tau \in [t_0, t_f]$ such that the following objective function is minimized:

$$J(u, x_0, t_0, t_f) = \int_{t_0}^{t_f} \left[\overbrace{x^T(t)Q(t)x(t)}^{\substack{\text{penalizes the transient} \\ \text{state deviation}}} + \overbrace{u^T R(t)u(t)}^{\text{penalizes control effort}} \right] dt + \underbrace{x(t_f)^T S x(t_f)}_{\text{penalizes the finite state}}$$

$$(4.23)$$

where $Q(t)$ and S are symmetric positive semi-definite $n \times n$ matrices, $R(t)$ is a symmetric positive definite $m \times m$ matrix. Notice that x_0, t_0, and t_f are fixed and given data (see Table 2.2 "Properties for the theorem of Lyapunov" in Chapter 2). The control goal is generally to keep $x(t)$ close to 0, especially, at the final time t_f, using little control effort u. That is to say, note in (4.23): this formulation can accommodate regulating an output

$$y = Cx \in \mathfrak{R}^r \qquad (4.24)$$

at near 0. In this case, one choice for S and $Q(t)$ are $C^T(t)W(t)C(t)$ where $W(t) \in \mathfrak{R}^{r \times r}$ is symmetric positive definite matrix [26].

4.2.4 Observer Based Controllers

In this section we deal with the design of observers for linear controllers. The usual strategy to obtain a good performance is to choose a very high observer gain. Using an observer with high gain results in a very fast decay of the estimation error, such that the system behaves as it would under state feedback after a short time span and thus reaches approximately the same performance as understate feedback. Unfortunately, a system with a very fast observer is sensitive to noisy measurements and modeling inaccuracies. If on the other hand a lower observer gain is chosen to a void these issues, there is no guarantee that the performance will be acceptable. It is necessary to consider that using the maximum input during a longer time span might damage the actuator [27].

Consider the following LTI system with a linear feedback control

$$\Sigma \begin{cases} \dot{x} = Ax + Bu \\ y = Cx \\ y_M = C_M x \\ u = -Kx \end{cases} \qquad (4.25)$$

Here $x = \Re^n$, the plant input $u = \Re^p$ and the output y_M we denote the vector of measured outputs $|u_i| \le u_{max,i}$ $\forall i = 1, ..., p$. The absolute value $|u_i|$ of each input is constrained by $u_{max,i}$, X_0 is assumed to be a convex subset of the null controllable region C.

To allow the stabilization of Σ using measurement feedback, the system is extended by a Luenberger observer to

$$\Gamma \begin{cases} \dot{x} = Ax + Bu \\ \dot{\hat{x}} = Ax + Bu + LC_M(x - \widetilde{x}) \\ u = -K\widetilde{x} \end{cases} \tag{4.26}$$

where x is the state of the plant, $\widetilde{x}$ the state of the observer, and L the observer feedback matrix. For ease of exposition a full observer is chosen, but the results can readily be extended to reduced observers. The composite system Γ can be written in the coordinates x and e as the familiar equations

$$\dot{x} = (A - BK)x + BKe \tag{4.27}$$

$$\dot{e} = (A - LC_M)e \tag{4.28}$$

where $e = x - \widetilde{x}$ is the observation error [28].

4.3 Nonlinear Controllers

Consider the following problem:

$$\begin{matrix} \dot{x} = f(x, u) \\ y = h(x) \end{matrix} \quad \textit{find} \quad \begin{rcases} u = \rho(x) \\ u = P(y) \end{rcases} \quad \begin{matrix} state - feeback \\ output - feedback \end{matrix} \tag{4.29}$$

so that the closed loop system $\dot{x} = f(x, \rho(x))$ or $\dot{x} = f(x, P(y))$ exhibits the desired stability (x is bounded and goes to 0) and performance (how it goes to set-point) characteristics $x \in \Re^n$ and $u \in \Re^m$, ρ and P may include dynamics [29].

Why do we use nonlinear control?

- To diminish the sensitivity to plant parameters.
- Tracking, regulate state, state set-point.
- To guarantee the appropriate transients.
- To guarantee the desired stability properties.

In general

- *Linear controller.* $\dot{x} = Ax + By$, $u = Cx$
- *Linear dynamics, static nonlinearity.* $\dot{x} = Ax + By$, $u = c(x)$
- *Nonlinear dynamical controller.* $\dot{x} = f(x, y)$, $u = c(x)$

4.3.1 Nonlinear Controllability

Consider the nonlinear SISO system

$$\dot{x} = f(x) + g(x)u; \quad x \in \Re^n, u \in \Re^1$$

$$y = h(x), \quad y \in \Re \tag{4.30}$$

The derivative of the output

$$\dot{y} = \frac{dh}{dx}\dot{x} = \frac{dh}{dx}(f(x) + g(x)u) \equiv L_f h(x) + L_g h(x)u \tag{4.31}$$

where $L_f h(x)$ and $L_g h(x)$ are the Lie derivatives "$L_f h$ is the derivative of h along the vector field of $\dot{x} = f(x)$".

Repeated derivatives

$$L_f h(x) = \frac{d(L_f^{k-1}h)}{dx}f(x), L_g L_f h(x) = \frac{d(L_f h)}{dx}g(x) \tag{4.32}$$

The relative degree p of a system is defined as the number of integrators between the input and the output (the number of times y must be differentiated for the input u to appear).

A nonlinear system has a relative degree p if

$$L_g L_f^{i-1}h(x) = 0, i = 1, \ldots, p-1; L_g L_f^{p-1}h(x) \neq 0 \forall x \in D \tag{4.33}$$

Definition 4.13 *(Lie Brackets)*

$$[X\,Y](y) = Y'(y)X(y) - X(y)Y(y) \tag{4.34}$$

Iterated Lie brackets: $[X, [X, Y]], [[Y, X], [X, [X, Y]]]$, etc. Why are Lie brackets natural objects for controllability issues? For simplicity, from now on we assume that

$$f(x, u) = f_0(x) + \sum_{i=1}^{m} u_i f_i(x) \tag{4.35}$$

We consider the control system $\dot{x} = f(x, u)$ where the state is $x \in \Re^n$ and the control is $u \in \Re^m$. Let us assume that $f(x_e, u_e) = 0$

Theorem 4.8 *Local Controllability*

Let O be a non-empty open subset of $\Re^n$ and let $x_e \in O$. Let us assume that, for some $f_1, \ldots, f_m : O \to \Re^n$,

$$f(x, u) = \sum_{i=1}^{m} u_i f_i(x) \quad \forall(x, u) \in O \times \Re^n \tag{4.36}$$

Let us also assume that

$$\{h(x_e)h \in Lie\{f_1 \ldots f_m\}\} = \Re^n \tag{4.37}$$

Then the control system $\dot{x} = f(x, u)$ is small-time locally controllable at $(x_e, 0) \in \Re^n \times \Re^m$

Theorem 4.9 *Global Controllability*

Let O be a non-empty open subset of $\Re^n$ Let us assume that, for some $f_1, \ldots, f_m : O \to \Re^n$,

$$f(x, u) = \sum_{i=1}^{m} u_i f_i(x) \quad \forall(x, u) \in O \times \Re^n \tag{4.38}$$

Let us also assume that

$$\{h(x_e); h \in Lie\{f_1 \ldots f_m\}\} = \mathcal{R}^n \; \forall y \in O \tag{4.39}$$

Then, for every $(x^0, x^1) \in O \times O$ and for every $T > O$, there exists u belonging to $L^\infty((0, T); R_m)$ such that the solution of the Cauchy problem $\dot{x} = f(x, u(t))$, $x(0) = x^0$ satisfies $x(t) \in O, t \in [0, T]$ and $x(T) = x^1$.

Definition 4.14 *Lie Algebra Rank Condition*
We consider the control affine system $\dot{x} = f_0(x) + \sum_{i=1}^{m} u_i f_i(x)$ with $f_0 = 0$. One says that this control system satisfies the Lie algebra rank condition at $0 \in \mathcal{R}^n$ if

$$\{h(0); h \in Lie\{f_1 \ldots f_m\}\} = \mathcal{R}^n \tag{4.40}$$

One has the following theorem [30].

Theorem 4.10 If the f_i's are analytic in a neighborhood of $0 \in \mathcal{R}^n$ and if the control system is small-time locally controllable at $(0, 0) \in \mathcal{R}^n \times \mathcal{R}^m$, then this control system satisfies the Lie algebra rank condition at $0 \in \mathcal{R}^n$.

4.3.2 Exact Feedback Linearization

Consider the nonlinear control-affine system

$$\dot{x} = f(x) + g(x)u \tag{4.41}$$

Note: use a state feedback controller $u(x)$ to make the system linear.
A nonlinear state transformation $z = \Gamma(x)$ with Γ invertible for x in the domain of interest, Γ and Γ^{-1} continuously differentiable, is called a diffeomorphism [31].

Definition 4.15 *A Nonlinear System*

$$\dot{x} = f(x) + g(x)u \tag{4.42}$$

is feedback linearizable if there exists a diffeomorphism Γ whose domain contains the origin and transforms the system into the form

$$\dot{x} = Ax + B\varphi(x)(u - \alpha(x)) \tag{4.43}$$

with (A, B) controllable and $\varphi(x)$ non-singular for all x in the domain of interest.

4.3.3 Input-Output Linearization

Use state feedback $u(x)$ to make the control-affine system

$$\dot{x} = f(x) + g(x)u$$
$$y = h(x) \tag{4.44}$$

linear from the input v to the output y, the general idea: differentiate the output, $y = h(x)$, p times until the control u appears explicitly in y^p, and then determine u so that

$$y^p = v \qquad (4.45)$$

Consider a system (4.44) and nth order SISO with relative degree p from Differentiating the output

$$\dot{y} = \frac{dh}{dx}\dot{x} = L_f h(x) + \underbrace{L_g h(x)u}_{=0}$$

$$\vdots \qquad (4.46)$$

$$y^p = L_f^p h(x) + L_g L_f^{p-1} h(x)u$$

and hence the state feedback controller

$$u = (-L_f^p h(x) + v)\,(L_g L_f^{p-1} h(x))^{-1},\ y^p = v \qquad (4.47)$$

Notes

- A system with unstable zero dynamics is called the non-minimum phase.
- The dynamics of the $n - p$ states not observable in the linearized dynamics of y are called the zero dynamics.
- The order of the linearized system is p, corresponding to the relative degree of the system.
- Consequently, if $p < n$ then $n - p$ states are unobservable in y [32].

4.3.3.1 Lyapunov-Based Control Design Methods

Consider a system

$$\dot{x} = f(x, u) \qquad (4.48)$$

- Find stabilizing state feedback $u = u(x)$.
- Verify stability through Control Lyapunov function.
- Methods depend on structure of f.

4.3.3.2 Back-Stepping Control

We want to design a state feedback $u = u(x)$ that stabilizes

$$\dot{x}_1 = f(x_1) + g(x_1)x_2$$
$$\dot{x}_2 = u \qquad (4.49)$$

at $x = 0$ with $f(0) = 0$

Idea: see the system as a cascade connection. Design a controller first for the inner loop and then for the outer.

Suppose the partial system

$$\dot{x}_1 = f(x_1) + g(x_1)v \qquad (4.50)$$

can be stabilized by $v = \varphi(x_1)$ and there exists a Lyapunov function $V_1 = V_1(x_1)$ such that

$$\dot{V}_1(x_1) = \frac{dV_1}{dx_1}(f(x_1) + g(x_1)\varphi(x_1)) \le -Q(x_1) \qquad (4.51)$$

for some positive definite function Q. This is a critical assumption in backstepping control.

4.3.4 Nonlinear Sliding Mode

The sliding mode control (SMC) makes the system slide into the sliding surface, and thus into the sliding state [33–40]. SMC is a nonlinear control technique featuring remarkable properties of accuracy, robustness, and easy tuning and implementation, for example:

- When the system goes into the sliding state, the sliding mode variable is also able to converge to zero.
- In bioprocess applications, the error between control objectives and the reference input and its derivative to form the sliding surface is commonly used.
- The dynamic behavior of the system may be tailored by the particular choice of the sliding function.
- The closed loop response becomes totally insensitive to some particular uncertainties.
- The design of the SMC is composed mainly of two parts, the sliding surface and control law.

Consider the system (4.30), as mentioned, SMC synthesis entails two phases.

4.3.4.1 Sliding Surface Design
The first phase is the definition of a certain scalar function of the system state, say

$$\sigma(x): \mathfrak{R}^n \to \mathfrak{R} \tag{4.52}$$

Often, the sliding surface depends on the tracking error e_y together with a certain number of its derivatives. The function σ should be selected in such a way that its vanishing, $\sigma = 0$, gives rise to a "stable" differential equation any solution e_y of which will tend to zero eventually.

$$\sigma = \sigma\left(e, \dot{e}, \dots, e^{(k)}\right) \tag{4.53}$$

The most typical choice for the sliding manifold is a linear combination of the following type

$$\sigma = \dot{e} + c_0 e$$
$$\sigma = \ddot{e} + c_1 \dot{e} + c_0 e$$
$$\sigma = e^{(k)} + \sum_{i=0}^{k-1} c_1 e^{(i)} \tag{4.54}$$

The number of derivatives to be included (the "k" coefficient in (4.54)) should be $k = r - 1$, where r is the input output relative degree of (4.30). With properly selected c_i coefficients, if one steers the σ variable to zero, the exponential vanishing of the error and its derivatives is obtained. The trajectories of the controlled system are forced onto the sliding surface, along which the system behavior meets the design specifications [41, 42]. A typical form for the sliding surface is the following, which depends on just a single scalar parameter p

$$\sigma = \left(\frac{d}{dt} + p\right)^k$$
$$\therefore k = 1 \quad \sigma = \dot{e} + pe$$
$$\therefore k = 2 \quad \sigma = \ddot{e} + 2p\dot{e} + p^2 e \tag{4.55}$$

Notes:

- The choice of the positive parameter p is almost arbitrary, and defines the unique pole of the resulting "reduced dynamics" of the system when sliding.
- The integer parameter k is in contrast rather critical, it must be equal to $r - 1$, with r being the relative degree between y and u.
- This means that the relative degree of the σ variable is one.

4.3.4.2 Control Law First-Order Sliding Mode Control

The consecutive phase is discovering a control action that steers the system trajectories onto the sliding manifold, that is, in other words, the control is able to steer the σ variable to zero in finite time [43].

There are several approaches based on the SMC approach:

- standard (or first-order) SMC
- high-order SMC.

The control is discontinuous across the manifold $\sigma = 0$

$$u = -U \, sgn(\sigma) \tag{4.56}$$

that is

$$u = \begin{cases} -U & \sigma > 0 \\ U & \sigma < 0 \end{cases} \tag{4.57}$$

U is a sufficiently large positive constant. In order to solve the above problem (referred to as "chattering singularity") approximate (smoothed) implementations of SMC techniques have been suggested where the discontinuous "sign" term is replaced by a continuous smooth approximation.

$$\begin{array}{llll} SAT & u = -Usat(\sigma; \varepsilon) \equiv -U\frac{\sigma}{|\sigma|+\varepsilon} & \varepsilon > 0 & \varepsilon \approx 0 \\ TANH & u = -Utanh(\sigma/\varepsilon) & \varepsilon > 0 & \varepsilon \approx 0 \end{array} \tag{4.58}$$

Lyapunov Method The Lyapunov method provides a dynamic system stable analysis approach. If we can construct a positive definite derivative function, the system is stable. This kind of method is usually used to design sliding mode controllers [44].

Suppose the positive definite Lyapunov function is selected as

$$V = \frac{1}{2}\sigma^2 \tag{4.59}$$

thus,

$$\dot{V} = \sigma\dot{\sigma} \tag{4.60}$$

so, the convergent condition of sliding mode variable is

$$\sigma\dot{\sigma} < 0 \tag{4.61}$$

Eq. (1.8) expresses the convergence condition of a sliding mode surface

$$\sigma = -k_1\sigma - k_2 sgn(\sigma) \tag{4.62}$$

Notes

- If the control law satisfies the above conditions, it guarantees $\sigma\dot{\sigma} < 0$.
- The convergence speed is decided by positive constant k_1.
- Equation (4.62) is a nonlinear SMC law, which lets the sliding mode variable converge into the origin within finite time.

4.3.4.3 Control Law Second-Order Sliding Mode Control

Using the above described smooth approximations, some problems are attenuated, at the price of a loss of robustness. Second-order SMC algorithms are powerful alternatives that completely solve the chattering issue also without compromising the robustness properties.

Considering a state equation of single input nonlinear system as

$$\dot{x} = f(x) + g(x)u$$

$$y = \sigma(x) \tag{4.63}$$

where, $x \in \mathfrak{R}^n$ is system state variable, y is the output, u is the control input. Here, $f(x)$, $g(x)$ and $\sigma(x)$ are smooth functions on time. The control objective is making the output function $\sigma \equiv 0$.

If the relative degree $r = 1$, traditional SMC can be used to design the controller. Otherwise, it needs to design a second-order sliding mode controller. A second-order SMC is able to eliminate chattering [45].

$$\dot{\sigma} = \frac{\partial}{\partial t}\sigma(x) + \frac{\partial}{\partial t}\sigma(x)[f(x) + g(x)u] \tag{4.64}$$

$$\ddot{\sigma} = \underbrace{\frac{\partial}{\partial t}\dot{\sigma}(x) + \frac{\partial}{\partial t}\dot{\sigma}(x)[f(x) + g(x)u]}_{\phi(x)} + \underbrace{\frac{\partial}{\partial u}\dot{\sigma}(x)\dot{u}}_{\lambda(x)} \tag{4.65}$$

$$\ddot{\sigma} = \phi(x) + \lambda(x) \tag{4.66}$$

Definition 4.16 Suppose a given sliding mode variable is $\sigma(x)$, the second-order sliding mode surface (or the sliding mode manifold) is defined as

$$\sigma^2 = \{x \in X | \sigma(x) = \sigma^2 = 0\} \quad x \in \mathfrak{R}^n \tag{4.67}$$

To illustrate the tightness of the control algorithm, the following conditions must be met

- u is a continuous and bounded function
- $\|f(x)\|_2$ and $2\|g(x)\|_2$, are bounded and $\gamma(x) > 0$

In meeting the above bounded conditions, there must be positive constants T_m, T_M, and Θ so that

$$0 < T_M \le \gamma(x) \le T_M$$

$$|\phi(x)| \le \Theta \tag{4.68}$$

4.3.4.4 Twisting Algorithm

According to knowledge of the nonlinear system coordinate transformation, new local coordinates $z_1 = \sigma$ and $z_2 = \dot{\sigma}$ are supposed [46]. Then, the original system (4.66) can be rewritten as

$$\dot{z}_1 = z_2$$
$$\dot{z}_2 = \varphi(x) + \lambda(x)\dot{u} \tag{4.69}$$

then

$$\dot{u} = \begin{cases} -u & |u| > u_0 \\ -V_m \, sgn(\sigma) & \sigma\dot{\sigma} > 0, |u| < u_0 \\ -V_M \, sgn(\sigma) & \sigma\dot{\sigma} \leq 0, |u| < u_0 \end{cases} \tag{4.70}$$

where, u_0, V_M, and V_m are positive constant. Similarly, if the systems relative degree is 2, the control algorithm is

$$u = \begin{cases} -V_m \, sgn(\sigma) & \sigma\dot{\sigma} \leq 0 \\ -V_M \, sgn(\sigma) & \sigma\dot{\sigma} > 0 \end{cases} \tag{4.71}$$

4.3.4.5 Super Twisting Algorithm

Only its control algorithm is continuous. Control input u is composed of two parts. One is differentiation of the discontinuous item [47]. The other is continuous functions of the sliding mode variable

$$u = u_1 + u_2 \tag{4.72}$$

$$\dot{u}_1 = \begin{cases} -u & |u| > 1 \\ -W \, sgn(\sigma) & |u| \leq 1 \end{cases} \tag{4.73}$$

$$u_2 = \begin{cases} -\lambda|\sigma_0|^P \, sgn(\sigma) & |\sigma| > \sigma_0 \\ -V_M|\sigma|^P \, sgn(\sigma) & |\sigma| \leq \sigma_0 \end{cases} \tag{4.74}$$

In addition, $p = 1$, the super twisting algorithm is convergent to the origin exponentially. At the moment, control input u hasn't the requirement of being bound. When $\sigma_0 = \infty$, the algorithm can be simplified as

$$u = u_1 - \lambda|\sigma|^P \, sgn(\sigma) \tag{4.75}$$

$$\dot{u}_1 = -W \, sgn(\sigma) \tag{4.76}$$

4.3.4.6 Variable Structure Systems

In the theory of variable structure systems (VSS) and variable structure control (VSC) the technique provides an easy way to design the control law or a plant. Numerous papers have been published in this area and their results are well documented [48–50]. The issues studied in detail include:

- Stability of the sliding mode.
- Existence of the sliding mode.
- Systems with time-varying coefficients.
- Unmeasurable linear or nonlinear state variables.
- Effects of system parameter perturbations.

In the next example a second-order system is considered using the following equation as has already been mentioned, the switching surface completely determines the plant dynamics in the sliding mode. Consequently, selecting this surface is one of the two major tasks in the process of SMC system design.

Statement of the VSC Problem equation For a given control system represented by the state

$$\dot{x} = A(x) + B(x)u \tag{4.77}$$

where $dim - x = n$ and $dim - u = m$, find: m switching functions, represented in vector form as $\sigma(x)$, and a VSC such that the reaching modes satisfy the reaching condition, namely, reach the set $\sigma(x) = 0$ (switching surface) in finite time.

$$u(x) = \begin{array}{ll} u^+(x) & \sigma(x) > 0 \\ u^-(x) & \sigma(x) < 0 \end{array} \tag{4.78}$$

The physical meaning of the above statement is as follows:

- Design a switching surface $\sigma(x) = 0$ to represent a desired system dynamics, which is of lower order than the given plant.
- Design a VSC $u(x)$ such that any state x outside the switching surface reaches the surface in finite time.

The most general state equation for nonlinear systems dynamics is

$$\dot{x} = f(x, u) \tag{4.79}$$

The most commonly used special form of (4.77) describes systems that are linear in the input u. Note that some systems that are not linear in the input u can still be put in the form of (4.77) by using an invertible term is represented by appending an m-dimensional output function

$$y = c(x, u) \tag{4.80}$$

Three points are stressed at the outset for nonlinear systems:

- Basic concepts and the fundamental theory of VSC are similar to those for linear systems.
- The derivation of a control $u(x)$ is similarly simple and straightforward.
- But the analysis of the sliding mode and the search for the corresponding switching function become a more difficult problem.

To study the stability of sliding modes occurring in nonlinear systems, various state transformations are used to put the differential equations of the system in one of several possible canonical forms. Reaching laws are generally tailored to take advantage of canonical form characteristics.

Fillipov Control Approach Fillipov established a systematic mathematical theory for differential equations with discontinuities [51]. Consider a general plant (4.77), given that the system is of variable structure, the system dynamics can be described by two structures

$$f(x) = \begin{array}{ll} f^+(x) & \sigma(x) > 0 \\ f^-(x) & \sigma(x) < 0 \end{array} \tag{4.81}$$

The system dynamics are not directly defined on $\sigma(x) = 0$ by (4.76). Instead, Fillipov describes the dynamics on $\sigma(x) = 0$ as a type of "average" of the two structures in (4.76)

$$\dot{x} = f_0(x) = uf_0^+ + (1 - u)f_0^- \tag{4.82}$$

The term u is also a function of x and can be specified in such a way that the "average" dynamic $f_0(x)$ is tangential to the surface $\sigma(x) = 0$.

Equivalent Control Approach The equivalent control approach is to find the input $u_{ev}(x)$ such that the state trajectory stays on the switching surface $\sigma(x) = 0$. Once the equivalent control input is known, the sliding mode dynamics can be described by substituting $u_{ev}(x)$ in (4.72). The equivalent control is found by recognizing that $\dot{\sigma}(x) = 0$ is a necessary condition for the state trajectory to stay on the switching surface $\sigma(x) = 0$. Differentiating $\sigma(x)$ with respect to time along the trajectory of (4.72) gives

$$\dot{\sigma}(x) = \frac{\partial \sigma}{\partial x} A(x) + \frac{\partial \sigma}{\partial x} B(x) u = 0 \tag{4.83}$$

Solving (4.84) for u yields the equivalent control

$$u_{ev}(x) = -\frac{\partial \sigma}{\partial x} A(x) \left[\frac{\partial \sigma}{\partial x} B(x) \right]^{-1} = 0 \tag{4.84}$$

where the existence of the matrix inverse is a necessary condition [52–54].

4.3.5 Model Predictive Control

MPC is formulated as the repeated solution of a (finite) horizon open-loop optimal control problem subject to system dynamics and input and state constraints. The MPC method is based on the receding horizon technique. The predictions are used by a numerical optimization program to determine the control signal that minimizes the following performance criterion over the specified horizon. Since bioprocesses are highly nonlinear systems, likely the application of nonlinear MPC methods using nonlinear/dynamic models becomes necessary. Nonlinear MPC strategies have already been successfully applied to bioprocesses [55–60].

The key characteristics and properties of nonlinear model predective control (NMPC) are:

- NMPC allows the direct use of nonlinear models for prediction.
- NMPC allows the explicit consideration of state and input constraints.
- In NMPC a specified time domain performance criteria is minimized on-line.
- In NMPC the predicted behavior is, in general, different from the closed loop behavior.
- For the application of NMPC typically a real-time solution of an open-loop optimal control problem is necessary.
- To perform the prediction the system states must be measured or estimated.
- Obtain estimates of the states of the system (depends on process model).
- Calculate an optimal input minimizing the desired cost function over the prediction horizon using the system model for prediction.
- Implement the first part of the optimal input until the next sampling instant.

However, the inputs will not be optimal in the sense that a deviation between the model and the plant might occur and since the controller does not recognize it, the computed control action will not minimize this deviation (referred to as model-plant mismatch).

Consider the class of continuous time systems described by the following nonlinear differential equation

$$\dot{x} = f(x, u), \quad x(0) = x_0 \tag{4.85}$$

subject to input and state constraints of the form:

$$
\begin{aligned}
u &\in \Theta, \forall \quad & t \geq 0 \\
x &\in X, \forall \quad & t \leq 0
\end{aligned} \tag{4.86}
$$

Furthermore, the input constraint set Θ is assumed to be compact and X is connected. For example Θ and X are often given by box constraints of the form:

$$\Theta := \{u \in \mathfrak{R}^m | u_{min} \leq u \leq u_{max}\} \text{ and } X := \{x \in \mathfrak{R}^n | x_{min} \leq x \leq x_{max}\} \tag{4.87}$$

In NMPC the input applied to the system is usually given by the solution of the following finite horizon open-loop optimal control problem, which is solved at every sampling instant:

$$min_{\overline{u}} = J(x, \overline{u}(\bullet)) \tag{4.88}$$

with the cost functional

$$J(x, \overline{u}(\bullet)) = \int_t^{t+T_p} F(\overline{x}, \overline{u}) d\tau \tag{4.89}$$

Here T_p and T_c are the prediction and the control horizon with $T_c \leq T_p$. The bar denotes internal controller variables and is the solution of the input signal $\overline{u}(\bullet)$ under the initial condition x.

The cost functional J is defined in terms of the stage cost F, which specifies the performance. The stage cost can for example arise from economical and ecological considerations.

$$F(x, u) = (x - x_s)^T Q(x - x_s) + (u - u_s)^T R(u - u_s) \tag{4.90}$$

Here x_s and u_s denote the desired reference trajectory, that can be constant or time-varying. The deviation from the desired values is weighted by the positive definite matrices Q and R. The value function plays a central role in the stability analysis of NMPC, since it often serves as a Lyapunov function candidate.

4.3.6 Control Using Neural Network

The inherent nonlinearity of the fermentation process often renders control difficult. Neural networks have become popular tools for the modeling and control of dynamic process, demonstrating the ability of handling nonlinearity. Many neural network controllers are of the rule-based type where the controller's output response is described by a series of control rules.

The unique features of this neural network control technique include:

- A wide operation range for handling a nonlinear process.
- Robustness for dealing with random disturbance and possible system parameter drafting.
- Relatively simple implementation.

The neural network predictive controller that is implemented in the neural network toolbox software uses a neural network model of a nonlinear plant to predict future plant performance. The controller then calculates the control input that will optimize plant performance over a specified future time horizon [61–65].

4.3.7 Nonlinear Design of Adaptive Controllers

In general, the dynamics of the bio-systems containnonlinearities and if the control scheme is based on feedback linearization, the problem of inaccurate modeling is often encountered. This is an important problem, since even very small differences between the real bio-system (experimental phenomenal) and the mathematical model may lead to large output errors. It is possible to determine the bounds on the system uncertainties and design a robust control approach, which guarantees stable control of the system as long as the uncertainties stay within these bounds. However, robust control does not enhance its performance with time. The solution to this problem is the use of adaptive control methods. Most of the work published so far relies on the availability of the full state vector for feedback by adaptive controllers [66–70].

Adaptive control tries to remove the effects of the uncertainties in the system model. Let $\pi \in \mathfrak{R}^p$ be a vector of unknown parameters. Since π is unknown, an estimate of it is used in the mathematical model of the biological system. If the vector of the estimated parameters is denoted by $\hat{\pi} \in \mathfrak{R}^p$, the ideal condition is that $\hat{\pi} = \pi$.

The objective of the adaptive control is to provide a mechanism such that the model parameters converge to the experimental values, even if the actual bio-systems parameters change unexpectedly.

4.3.7.1 Identification of Unknown Parameters

$$\dot{x} = f(x, \pi) + g(x, \pi)u \tag{4.91}$$

where $x \in \mathfrak{R}^n$ is the state vector, $f, g : \Omega \subset \mathfrak{R}^n \to \mathfrak{R}^n$ are sufficiently smooth vector fields, $\pi \in \Xi \subset \mathfrak{R}^p$ is the unknown parameter vector, and $u \in \mathfrak{R}$ is the input. The vector fields f and g have the form,

$$f(x, \pi) = \sum_{k=1}^{p} \pi_k f_k(x) \tag{4.92}$$

$$g(x, \pi) = \sum_{k=1}^{p} \pi_k g_k(x) \tag{4.93}$$

The regressor $w(x, \pi) \in \mathfrak{R}^{p \times n}$ is formulated as,

$$w(x, \pi) = \begin{bmatrix} f_1(x) + g_1(x)u \\ \vdots \\ f_p(x) + g_p(x)u \end{bmatrix} \tag{4.94}$$

which includes all the nonlinearities in the system dynamics. Hence, we can write (4.91) in the regressor form as,

$$\dot{x} = w^T(x, u)\pi \tag{4.95}$$

4.3.7.2 Observer-Based Identification

The identifier system for the observer-based method is given with the following identification dynamics equations,

$$\dot{\hat{x}} = A(\hat{x} - x) + w^T(x, u)\hat{\pi}$$
$$\dot{\hat{\pi}} = -w(x, u)P(\hat{x} - x) \tag{4.96}$$

where $\hat{x} \in \mathfrak{R}^n$ and $\hat{\pi} \in \mathfrak{R}^p$ are estimates of x and π respectively, $A \in \mathfrak{R}^{n \times n}$ is a Hurwitz matrix and $P \in \mathfrak{R}^{n \times n}$ is the positive definite solution of the Lyapunov equation

$$A^T P - PA = -Q \tag{4.97}$$

with $Q \in \mathfrak{R}^{n \times n}$ being any symmetric, positive definite matrix. Defining $\tilde{x} = \hat{x} - x$ as the observer state error and $\tilde{\pi} = \hat{\pi} - \pi$ as the parameter error, we obtain the following error system:

$$\dot{\hat{x}} = A\tilde{x} + w^T(x, u)\tilde{\pi}$$
$$\dot{\tilde{\pi}} = -w(x, u)P\tilde{x} \tag{4.98}$$

Theorem 4.11   Let the identifier error dynamics of the system in (4.96) be given by (4.92). Then, $\tilde{\pi} \in \mathfrak{I}_\infty$ and $\tilde{x} \in \mathfrak{I}_2 \cap \mathfrak{I}_\infty$. If the system given in (4.96) is bounded-input bounded-state (BIBS) stable as well, then $\tilde{x} \in \mathfrak{I}_\infty$ and $\tilde{x} \to 0$ as $t \to \infty$ [71].

4.3.7.3 Adaptive Control Under Matching Conditions

The matching condition can be described as the case when the unknown parameters are in the same equation as the control input. This methodology is introduced by Taylor et al. [72], continued by Kanellakopoulos, Kokotovic, and Marino [73], and formalized by Pomet and Praly [74]. Consider the following nonlinear system:

$$\dot{x} = f(x, \pi) + g(x, \pi)u$$
$$y = h(x) \tag{4.99}$$

where $x \in \mathfrak{R}^n$ is the state vector, $u \in \mathfrak{R}$ is the input, $\pi \in \Xi \subset \mathfrak{R}^p$ is the vector of unknown parameters, $f, g : \Omega \subset \mathfrak{R}^{n \times p} \to \mathfrak{R}^n$ are sufficiently smooth vector fields, $h : \Psi \subset \mathfrak{R}^n \to \mathfrak{R}$ is a sufficiently smooth function, and $y \in \mathfrak{R}$ is the system output.

Since the parameter vector π is unknown, we use the estimate of the parameter vector and re-write the model as,

$$\dot{x} = f(x, \pi) - \Delta f(x) + g(x, \hat{\pi})u + \Delta g(x)u$$
$$y = h(x) \tag{4.100}$$

where $\hat{\pi}$ is the estimate of the parameter vector

$$\Delta f(x) = f(x, \pi) - f(x, \hat{\pi}) \tag{4.101}$$
$$\Delta g(x) = g(x, \pi) - g(x, \hat{\pi}) \tag{4.102}$$

In (4.100), $f(x, \hat{\pi})$ and $g(x, \hat{\pi})$ represent the known nominal parts because the state x is assumed to be measurable, and the vector of parameter estimates $\hat{\pi}$ is available. The uncertainty is considered in $\Delta g(x)$ and $\Delta f(x)$.

Proposition 4.3 Consider the system in (4.99) and the model in (4.100). If the relative degree of the system is γ, and if the uncertainties $\Delta g(x)$ and $\Delta f(x)$ satisfy then the linearizing control law for the system in (4.99) also linearizes the model in (4.100) [75].

$$L_{\Delta f} L^i_{f(x,\hat{\pi})} h = 0, \qquad i = 0, \ldots, \gamma - 1$$
$$L_{\Delta g} L^i_{f(x,\hat{\pi})} h = 0, \qquad i = 0, \ldots, \gamma - 1 \qquad (4.103)$$

then the linearizing control law for the system in (4.99) also linearizes the model in (4.100).

4.3.7.4 Indirect Adaptive Control

The nonlinear system is as given in (4.99). The indirect scheme estimates the plant parameters by incorporating a separate identifier model. Due to the system dynamics (4.101) can be written in the regressor and vector fields f and g have the form. The unknown parameter vector π in (4.101) can be identified using the filtering-based identification method. As a result, x_f and w_f being the filtered forms of x and w respectively, a gradient update rule is derived as,

$$\hat{\pi} = -\alpha w_f \tilde{x} \quad \forall \alpha \in \mathfrak{R}^+ \qquad (4.104)$$

where $\tilde{x} = \hat{x} - x$ is the state prediction error. The asymptotic convergence of $\tilde{x}$ to zero [71]. Furthermore, the convergence of parameter estimate vector $\hat{\pi}$ to the actual system parameter vector π is expressed under persistently exciting signal condition. With the estimated system parameters at hand, in order to generate the control signal based on these estimates, we write the system dynamics in (4.103) in the normal form where $\dot{v} = q(\xi, v)$ defines the internal dynamics with exponentially stable zero-dynamics

$$\dot{\xi}_1 = \xi_2$$
$$\vdots$$
$$\dot{\xi}_r = L^r_f h(x)(x, \pi) + L_g L^{r-1}_f h(x)(x, \pi)u$$
$$\dot{v} = q(\xi, v)$$
$$y = \xi_1 \qquad (4.105)$$

Theorem 4.12 Consider the nonlinear system defined in (4.93) with a bounded tracking signal y_r, with bounded derivatives as $y_r, \dot{y}_r, \ldots, y_r^\lambda$; and with the internal dynamics $v(\xi, \lambda)$ which is globally Lipschitz in ξ, λ. If the regressor vector φ_r is bounded for bounded ξ, λ, u such that $\|\varphi^T(\xi, \lambda, u)\| \le b_\varphi(\|\xi\| + \|\lambda\|)$ holds, and if the input is persistently exciting such that $\tilde{\Theta} \to 0$ as $t \to \infty$, then the control laws given by (4.106) and (4.105) along with the parameter update rule in (4.105) result in a closed loop stable system with tracking convergence such that $y(t) \to y_r(t)$ as $t \to \infty$ [76].

4.3.7.5 Model Reference Adaptive Control

Here we introduce a model reference adaptive controller (MRAC) scheme for a subset of the system defined in (4.85). Consider the following system,

$$\dot{x} = f(x) + g(x)u + \sum_{k=1}^p q_k(x)\pi_k \qquad (4.106)$$

here $x \in \mathfrak{R}^n$ is the state vector, $u \in \mathfrak{R}$ is the input, is the vector of unknown parameters, f, $g \in \mathfrak{R}^n$ are sufficiently smooth vector fields, with $f(0) = 0$, $g(0) = 0$, $\pi = [\pi_1, \ldots, \pi_p]^T$ is the constant but unknown parameter vector which takes values on a compact set $\pi \in \Xi \subset \mathfrak{R}^p$. $q_k \in \mathfrak{R}^n$ are known vector fields. We assume that the origin is the equilibrium point of (4.102), without loss of generality and $u \in \mathfrak{R}$ is the input. We assume that the origin is the equilibrium point of (4.102), without loss of generality. The nominal part (f, g) of the system in (4.102) is given by,

$$\dot{x} = f(x) + g(x)u \tag{4.107}$$

and is assumed to be input-state linearizable via state feedback. Since the input-state linearizability conditions stated, there exists a smooth scalar function $\wp(x) : \mathfrak{R}^n \to \mathfrak{R}$ with $\wp(0) = 0$ such that, $L_g L_f^i \wp(x) = 0$ $\forall i = 0, \ldots, n-2$ and $L_g L_f^{n-1} \wp(x) \neq 0$. Then we obtain the linearized state variables using the linearizing diffeomorphism $\xi = \varphi(x)$ as,

$$\xi_1 = L_f^{i-1} \wp(x), \forall i = 1, \ldots, n \tag{4.108}$$

The state feedback control law given by,

$$u = (-L_f^n \wp(x) + v) \; (L_g L_f^{p-1} \wp(x))^{-1}, y^p = v \tag{4.109}$$

converts the nominal system in (4.107) into a linear system in Brunowsky canonical form with the new state vector ξ as,

$$\dot{\xi}_1 = \xi_{i+1}$$
$$\dot{\xi}_n = v \tag{4.110}$$

which is equivalent to,

$$\dot{\xi} = A\xi + bv \tag{4.111}$$

where (A, b) is a controllable pair. Since the nominal system f, g in (4.107) is also input state linearizable, we can now implement a model reference adaptive controller for the original system.

In general, beneficial impacts of control theory (nonlinear systems) finally reached the bio-industry [77, 78]. However, there is as yet no effective standard for the methodologies to implementation (scale-up). In addition, three challenges have been identified that are critical for the realization and success of control theory framework: (i) Dynamic models are required that allow changes in the operating conditions and perturbations to be predicted. (ii) New design of experiments (DoE) approaches will lead to a reduced workload and/or better understanding of the bioprocess and its dynamics enabling the development of more meaningful and predictive dynamic models. (iii) Soft-sensors are required that provide estimations of the states to enable the dynamic model predictions and closing the control loop to be corrected. (iv) Flexible and new software frameworks will be essential, offering the ability to incorporate different types of data sets, at, on, and in-line, respectively, and using the built up model in real time [79] (see Table 4.2).

Table 4.2 Control techniques actually applied to real bioprocesses [29].

Control theory	Nonlinear dynamics	Unpredictable dynamics disturbances	Process model	Historical data set	Experience	Highlights
Open loop	■	■			■	An appropriate input wave shaping; simultaneous optimization method; the batch reactor is operated in open loop [80, 81].
DoE-SPC [82–85]	■			■	■	Dynamic hybrid model; feedforward artificial neural network; simulated *Escherichia coli* fermentation.
MPC [86–90]	■		■		■	These techniques are used in order to regulate the substrate concentration and the biomass production in a bioreactor, are applied to a continuous bioreactor in which the pH changes are considered as disturbances, process model.
Fuzzy [91–95]	■	■				Input multiplicities in dilution rate on productivity; controller designed to translate the information obtained from the operator's experiences; automatic control system.
Adaptive [62, 96–100]	■	■	■			Anaerobic bioreactor; modeling; adaptive proportional gain; robust performance; performance of algorithm NEPSAC (extended, predictive, and self-adaptive nonlinear controller); embedded system for the collect of the data.
ANN [91, 101–103]	■	■		■		Neural networks belong to data-driven models; applied in the biochemical processing industries; applied in modeling complex nonlinear processes whose process understanding is limited; data-based empirical model.
SMC [104–108]	■	■	■			High-order sliding mode controller has a great impact on chattering reduction; optimal transient operation of a continuous bioreactor; optimality and robustness properties; is applied through simulations to an anaerobic digester.

List of Figures

List of Tables

References

1 Behrooz, F., Mariun, N., Marhaban, M.H. et al. (2018). Review of control techniques for HVAC systems – nonlinearity approaches based on fuzzy cognitive maps. *Energies* 11: 495.

2 Malik, A., Meyr, U., Herfs, W. et al. (2018). Control of Dynamically Inherent Biological Processes in Cell Technology. IEEE International Conference on Engineering, Technology and Innovation (ICE/ITMC), Stuttgart, 2018.

3 Vojtesek, J. and Dostal, P. (2015). Nonlinear versus ordinary adaptive control of continuous stirred-tank reactor. *The Scientific World Journal* 1: 389273.

4 Liu, T., Wang, X.Z., and Chen, J. (2014). Robust PID based indirect-type iterative learning control for batch processes with time-varying uncertainties. *Journal of Process Control* 24 (12): 95–106.

5 Whalen, A.J., Brennan, S.N., Sauer, T.D. et al. (2015). Observability and controllability of nonlinear networks: the role of symmetry. *Physical Review* 5 (1): 011005.

6 Behrooz, F., Ramli, A.R., Samsudin, K. et al. (2017). Energy saving by applying the fuzzy cognitive map control in controlling the temperature and humidity of room. *International Journal of Physical Sciences* 12: 13–23.

7 Sommeregger, W., Sissolak, B., Kandra, K. et al. (2017). Quality by control: towards model predictive control of mammalian cell culture bioprocesses. *Biotechnology Journal* 12: 1600546.

8 Prieto-Escobar, N., Saldarriaga-Aristizábal, P.A., and Chaparro-Muñoz, V. (2018). Heuristic parameter estimation for a continuous fermentation bioprocess. *Revista Facultad de Ingeniería Universidad de Antioquia* 88: 26–39.

9 Wei, J. and Yu, Y. (2018). An effective hybrid cuckoo search algorithm for unknown parameters and time delays estimation of chaotic systems. *IEEE Access* 6: 6560–6571.

10 Pantano, M.N., Serrano, M.E., Fernández, M.C. et al. (2017). Multivariable control for tracking optimal profiles in a nonlinear fed-batch bioprocess integrated with state estimation. *Industrial & Engineering Chemistry Research* 56 (20): 6043–6056.

11 Mahdianfar, H. and Pavlov, A. (2017). Adaptive output regulation for offshore managed pressure drilling. *International Journal of Adaptive Control and Signal Processing* 31: 652–673.

12 Huang, Y. and Liu, Y. (2018). Adaptive output-feedback control of nonlinear systems with multiple uncertainties. *Asian Journal of Control* 20: 1151–1160.

13 Harmand, J., Rapaport, A., Dochain, D. et al. (2018). Microbial ecology and bioprocess control: opportunities and challenges. *Journal of Process Control* 18: 865–875.

14 Domınguez-Garcıa, J.L. and Garcıa-Planas M.I. (2011). *Output controllability and steady-output controllability analysis of fixed speed wind turbine.* Physcon, Léon, Spain, September, 5–September.

15 Hespanha, J.P. (2009). *Linear Systems Theory.* Princeton University Press.

16 Qiu, D., Wang, Q., and Zhou, Y. (2009). *Steady-State Output Controllability and Output Controllability of Linear Systems*, 147–150. Computational Intelligence and Industrial Applications, IEEE Explore.

17 Craven, S., Whelan, J., and Glennon, B. (2014). *Glucose concentration control of a fed-batch mammalian cell bioprocess using a nonlinear model predictive controller.* *Journal of Process Control* 24: 344–357.

18 Zamorano, F., Vande Wouwer, A., Jungers, R.M. et al. (2013). *Dynamic metabolic models of CHO cell cultures through minimal sets of elementary flux modes.* *Biotechnology Journal* 164: 409–422.

19 Villaverde, A.F., Barreiro, A., and Papachristodoulou, A. (2016). *Structural Identifiability of dynamic systems biology models.* *PLoS Computational Biology* 12: e1005153.

20 Duran-Villalobos, C.A., Lennox, B., and Lauri, D. (2016). *Multivariate batch to batch optimisation of fermentation processes incorporating validity constraints. Journal of Process Control* 46: 34–42.

21 Golabgir, A., Hoch, T., Zhariy, M. et al. (2015). *Observability analysis of biochemical process models as a valuable tool for the development of mechanistic soft sensors.* *Biotechnology Progress* 31 (6): 1703–1715.

22 Villaverde, A.F. and Banga, J.R. (2017). *Structural properties of dynamic systems biology models: identifiability, reachability, and initial conditions. Processes* 5: 29.

23 Mannakee, B.K., Ragsdale, A.P., Transtrum, M.K. et al. (2016). *Sloppiness and the geometry of parameter space.* In: *Uncertainty in Biology*, 271–299. Cham, Switzerland: Springer.

24 Balchen, J., Andresen, G.T., and Foss, B.A. (2004). *Reguleringsteknikk.* Institutt for teknisk kybernetikk.

25 Saez-Rodriguez, J. (2014). *MEIGO: an open-source software suite based on metaheuristics for global optimization in systems biology and bioinformatics. BMC Bioinformatics* 15: 136.

26 Liu, K. and Yao, Y. (2016). *Fundamentals of linear system. In robust control.* In: *Robust Control: Theory and Applications* (eds. K. Liu and Y. Yao). Wiley.

27 Lens, H. and Adamy, J. *Observer Based Controller Design for Linear Systems with Input Constraints* . Proceedings of the 17th World Congress The International Federation of Automatic Control Seoul, Korea, July 6–11, 2008.

28 Alamo, T., Cepeda, A., and Limon, D. (2005). *Improved computation of ellipsoidal invariant sets for saturated control systems.* In: *44th IEEE Conference on Decision and Control, 2005 and 2005 European Control Conference. CDC-ECC '05*, 6216–6221, 12–15 Dec. 2005.S.

29 Mears, L., Stocks, S.M., Sin, G. et al. (2017). *A review of control strategies for manipulating the feed rate in fed-batch fermentation processes. Journal of Biotechnology* 245: 34–46.

30 Maya-Yescas, R. and Aguilar, R. (2003). *Controllability assessment approach for chemical reactors: nonlinear control affine systems. Chemical Engineering Journal* 92: 69–79.

31 Cai, P., Wu, X., Sun, R. et al. (2017). *Exact feedback linearization of general four-level buck DC-DC converters*. In: *2017 29th Chinese Control and Decision Conference (CCDC)*, 4638–4643. Chongqing: IEEE.

32 Isidori, A. (2013). *The zero dynamics of a nonlinear system: from the origin to the latest progresses of a long successful story. European Journal of Control* 19 (5): 369–378.

33 Lara-Cisneros, G., Femat, R., and Dochain, D. (2017). *Robust sliding mode based extremum-seeking controller for reaction systems via uncertainty estimation approach. International Journal of Robust and Nonlinear Control* 27: 3218–3235.

34 Singla, M., Shieh, L., Song, S. et al. (2014). *A new optimal sliding mode controller design using scalar sign function. ISA Transactions* 53 (2): 267–279.

35 Cui, Y. and Xu, L. (2019). *Chattering-free adaptive sliding mode control for continuous-time systems with time-varying delay and process disturbance. International Journal of Robust Nonlinear Control* 29: 3389–3404.

36 Mezghani, N., Romdhane, B., and Damak, T. (2010). *Terminal sliding mode feedback linearization control. International Journal of Automation and Computing* 4: 1174–1187.

37 Xiang, W. and Huangpu, Y. (2010). Second-order terminal sliding mode controller for a class of chaotic systems with unmatched uncertainties. *Communications in Nonlinear Science and Numerical Simulation* 15: 3241–3247.

38 Skruch, P. and Długosz, M. (2019). *Design of terminal sliding mode controllers for disturbed non-linear systems described by matrix differential equations of the second and first orders. Applied Sciences* 9: 2325.

39 Bartoszewicz, A. and Nowacka, A. (2007a). *Sliding mode control of the third-order system subject to velocity, acceleration and input signal constraints. Proceedings of the IET Part D: Control Theory and Applications* 1: 1461–1470.

40 Kanthalakshmi, S. and Annal, W.P. (2018). *Real time implementation of adaptive sliding mode controller for a nonlinear system. Studies in Informatics and Control* 27 (4): 395–402.

41 Mahmoud, M.S. (2018). *Advanced Control Design with Application to Electromechanical Systems*, 1e. MA, USA: Butterworth-Heinemann Newton.

42 Meza-Aguilar, M., Loukianov, A.G., and Rivera, J. (2019). *Sliding mode adaptive control for a class of nonlinear time-varying systems. International Journal of Robust Nonlinear Control* 29: 766–778.

43 Castanos, F. and Fridman, L. (2006). *Analysis and design of integral sliding manifolds for systems with unmatched perturbations. IEEE Transactions on Automatic Control* 51: 853–858.

44 Sánchez, T. and Moreno, J.A. (2014). A constructive Lyapunov function design method for a class of homogeneous systems. In: *53rd IEEE Conference on Decision and Control*, 5500–5505. Los Angeles, CA: IEEE.

45 Yang, P., Fang, Y., Wu, Y. et al. (2017). *Fast smooth second-order sliding mode control for systems with additive colored noises. PLoS One* 12 (5): e0178455.

46 Derbeli, M., Barambones, O., and Ramos-Hernanz, et al. (2019). *Time implementation of a super twisting algorithm for PEM fuel cell power system. Energies* 12: 1594.

47 Figueroa-Estrada, J.C., Neria-González, M.I., and Aguilar-López, R. (2019). *Design of a class of super twisting sliding-mode controller: application to bioleaching process. Comptes rendus de l'Acade'mie bulgare des Sciences* 72 (7): 947–954.

48 Hung, J.Y., Gao, W., and Hung, J.C. (1993). *Variable structure control: a survey. IEEE Transactions on Industrial Electronics* 40 (1): 2–22.

49 Niu, H., Wang, Q., and Wang, H. (2013). *Variable structure control for methane fermentation systems.* In: *Proceedings of the 32nd Chinese Control Conference*, 866–870. Xi'an.

50 Rahman, A.F.N.A., Spurgeon, S.K., and Yan, X. (2010). *A sliding mode observer for estimating substrate consumption rate in a fermentation process.* In: *11th International Workshop on Variable Structure Systems*, 172–177. Mexico City: VSS.

51 Azhmyakov, V., Egerstedt, M., Fridman, L. et al. (2009). *Continuity properties of nonlinear affine control systems: applications to hybrid and sliding mode dynamics. IFAC Proceedings* 42 (17): 204–209.

52 Haskara, I. (1998). *On sliding mode observers via equivalent control approach. International Journal of Control* 71: 1051–1067.

53 Yasin, A.R., Ashraf, M., and Bhatti, A.I. (2019). *A novel filter extracted equivalent control based fixed frequency sliding mode approach for power electronic converters. Energies* 12: 853.

54 Pan, Y., Yang, C., Pan, L. et al. (2018). *Integral sliding mode control: performance, modification, and improvement. IEEE Transactions on Industrial Informatics* 14: 3087–3309.

55 Allgöwer, F., Findeisen, R., and Nagy, Z.K. (2004). *Nonlinear model predictive control: from theory to application. Journal of the Chinese Institute of Chemical Engineers* 35 (3): 299–315.

56 Degachi, H., Naffeti, B., Chagra, W. et al. (2018). *Filled function method for nonlinear model predictive control. Mathematical Problems in Engineering* 9497618: 1–8.

57 Faulwasser, T., Grüne, L., and Müller, M.A. (2018). Economic nonlinear model predictive control. *Foundations and Trends R in Systems and Control* 5 (1): 1–98.

58 Williams, G., Drews, P., and Goldfain, B.J. (2018). *Information-theoretic model predictive control: theory and applications to autonomous driving. IEEE Transactions on Robotics* 34: 1603–1622.

59 Lucia, S. and Engell, S. (2013). *Robust nonlinear model predictive control of a batch bioreactor using multi-stage stochastic programming.* In: *European Control Conference (ECC)*, 4124–4129. Zurich.

60 Bradford, E. and Imsland, L. (2017). *Expectation constrained stochastic nonlinear model predictive control of a batch bioreactor.* In: *Aided Chemical Engineering*, vol. 40 (eds. A. Espuña, M. Graells and L. Puigjaner), 1621–1626. Elsevier.

61 Mete, T., Ozkan, G., and Hapoglu, H. (2012). *Control of dissolved oxygen concentration using neural network in a batch bioreactor. Computer Applications in Engineering Education* 20: 619–628.

62 Imtiaz, U., Assadzadeh, A., and Jamuar, S.S. (2013). *Bioreactor temperature profile controller using inverse neural network (INN) for production of ethanol. Journal of Process Control* 23: 731–742.

63 Zhang, D., Chanona, A.E., and Del-Rio, V.S. (2015). *Analysis of green algal growth via dynamic model simulation and process optimisation. Biotechnology and Bioengineering* 112 (10): 2025–2039.

64 Machón-González, I. and López-García, H. (2017). *Feedforward nonlinear control using neural gas network. Complexity* 3125073: 11.

65 Vasičkaninová, A. and Bakošová, M. (2009). *Neural network predictive control of a chemical reactor. Acta Chimica Slovaca* 2 (2): 21–36.

66 Le, A.T. and Lee, S.-G. (2016). *Nonlinear feedback control of Underactuated mechanical systems.* In: *Nonlinear Systems – Design, Analysis, Estimation and Control* (eds. D. Lee, T. Burg and C. Volos), 243–263. IntechOpen.

67 Krstic, M., Kanellakopoulos, I., and Kokotovic, P.V. (1994). *Nonlinear design of adaptive controllers for linear systems. IEEE Transactions on Automatic Control* 39 (4): 738–752.

68 Roy, S. and Narayan, K.I. (2016). *Robust time-delayed control of a class of uncertain nonlinear systems. IFAC-Papers OnLine* 49: 736–741.

69 Bastin, G. and Van Impe, J.F. (1995). *Nonlinear and adaptive control in biotechnology: a tutorial. European Journal of Control* 1: 37–53.

70 Chitra, M., Pappa, N., and Abraham, A. (2018). *Dissolved oxygen control of batch bioreactor using model reference adaptive control scheme. IFAC-PapersOnLine* 51 (4): 13–18.

71 Andersen, P., Pedersen, T.S., and Nielsen, K.M. (2012). *Observer Based Model Identification of Heat Pumps in a Smart Grid*, 569–574. Dubrovnik: IEEE International Conference on Control Applications.

72 Taylor, D.G., Kokotovic, P.V., Marino, R. et al. (1989). *Adaptive regulation of nonlinear systems with Unmodeled dynamics. IEEE Transactions on Automatic Control* 34 (4): 405–412.

73 Kanellakopoulos, I., Kokotovic, P.V., and Marino, R. (1989). *Robustness of Adaptive Nonlinear Control under an Extended Matching Condition*, 192–197. Capri, Italy: IFAC Symposium on Nonlinear System Design.

74 Pomet, J.B. and Praly, L. (1992). *Adaptive nonlinear regulation: estimation from the Lyapunov equation. IEEE Transactions on Automatic Control* 37 (6): 729–740.

75 Zhang, M., Gan, M.G., Chen, J. et al. (2019). *Data-driven adaptive optimal control of linear uncertain systems with unknown jumping dynamics. Journal of the Franklin Institute* 356 (12): 6087–6105.

76 Cezayirli, A. and Ciliz, M.K. (2008). *Indirect adaptive control of nonlinear systems using multiple identification models and switching. International Journal of Control* 81 (9): 1434–1450.

77 Landau, I.D., Lozano, R., and M'Saad, M. (2011). *Adaptive Control. Communications and Control Engineering.* London: Springer-Verlag.

78 Amand, M.M.S., Tran, K., and Radhakrishnan, D. (2014). *Controllability analysis of protein glycosylation in cho cells. PLoS One* 9: 1–16.

79 Utkin, V.I. (1977). *Variable structure systems with sliding modes. IEEE Transaction on Automatic Control* 22 (2): 212–222.

80 Simutis, R. and Lübbert, A. (2015). *Bioreactor control improves bioprocess performance. Biotechnology Journal* 10: 1115–1130.

81 Jazayeri-Rad, H. and Shahbaznezhad, M. (2014). *Optimizing the open-and closed-loop operations of a batch reactor. Journal of Automation and Control Engineering* 2 (1): 8–19.

82 Stosch, M. and Willis, M.J. (2017). Intensified design of experiments for upstream bioreactors. *Engineering in Life Sciences* 17: 1173–1184.

83 Belmokhtar, F.Z., Elbahri, Z., and Elbahri, M. (2018). *Preparation and optimization of agrochemical 2,4-D controlled release microparticles using designs of experiments. Journal of the Mexican Chemical Society* 62 (1) 00005.

84 Hussein, H., Williams, D.J., and Liu, Y. (2015). *Design modification and optimisation of the perfusion system of a tri-axial bioreactor for tissue engineering. Bioprocess and Biosystems Engineering* 38: 1423.

85 Telen, D., Nimmegeers, P., and Van Impe, J. (2018). *Uncertainty in optimal experiment design: comparing an online versus offline approaches. IFAC-PapersOnLine* 51 (2): 771–776.

86 Ntampasi, V.E. and Kosmidou, O.I. (2015). *A comparison of robust model predictive control techniques for a continuous bioreactor.* In: *12th International Conference on Informatics in Control, Automation and Robotics*, 431–438. Colmar: ICINCO.

87 Jabarivelisdeh, B., Findeisen, R., and Waldherr, S. (2018). *Model predictive control of a fed-batch bioreactor based on dynamic metabolic-genetic network models. IFAC-PapersOnLine* 51 (19): 34–37.

88 Wu, Z., Tran, A.Y., Ren, M. et al. (2019). *Model predictive control of phthalic anhydride synthesis in a fixed-bed catalytic reactor via machine learning modeling. Chemical Engineering Research and Design* 145: 173–183.

89 Aadaleesan, P. and Saha, P.A. (2018). *Nash game approach to mixed H2/H∞ model predictive control: part 3 – output feedback case. International Journal of Automation and Computing* 15: 616.

90 Violaro, F., Rivera, E., and Alvarez, L. (2018). *Multivariable model predictive control of a continuous fermentation unit for first-generation ethanol production. Chemical Engineering Transactions* 65: 67–72.

91 Arulmozhi, N. (2014). *Bioreactor control using fuzzy logic controllers. Applied Mechanics and Materials* 573: 291–296.

92 Burdge, D.A. and Libourel, I.G.L. (2014). *Open source software to control Bioflo bioreactors. PLoS One* 9 (3): e92108.

93 Hussein, T. and Shamekh, A. (2019). *Design of PI fuzzy logic gain scheduling load frequency control in two-area power systems. Designs* 3: 26.

94 Tayyebi, S., Shahrokhi, M., and Bozorgmehry Boozarjomehry, R. (2010). *Fault diagnosis in a yeast fermentation bioreactor by genetic fuzzy system. Iranian journal of chemistry and chemical engineering* 29 (3): 61–72.

95 Patnaik, P.R. (1997). *Artificial intelligence as a tool for automatic state estimation and control of bioreactors. Laboratory Robotics and Automation* 9: 297–304.

96 Aguilar-López, R. (2018). *Chaos suppression via Euler-Lagrange control Design for a Class of chemical reacting system. Mathematical Problems in Engineering* 6: 1–6.

97 Aguilar-López, R., Neria-González, M.I., Martínez-Guerra, R. et al. (2014). *Nonlinear estimation in a class of gene transcription process. Applied Mathematics and Computation* 226: 131–144.

98 Aguilar-López, R. and Neria-González, I. (2016). *Controlling continuous bioreactor via nonlinear feedback: modelling and simulations approach. Bulletin of the Polish Academy of Sciences Technical Sciences* 64 (1): 235–241.

99 Peña-Caballero, V., López-Pérez, P.A., Neria-González, M.I. et al. (2012). *A class of nonlinear adaptive controller for a continuous anaerobic bioreactor. Journal of Scientific and Industrial Research* 71 (07): 480–486.

100 Pablo, P.R., William, I.A., Jose, M. et al. (2016). Predictive and adaptive nonlinear controller applied to a drying process of cocoa beans. IEEE Ecuador Technical Chapters Meeting, ETCM.

101 Wong, W.C., Chee, E., Li, J. et al. (2018). *Recurrent neural network-based model predictive control for continuous pharmaceutical manufacturing. Mathematics* 6: 242.

102 Rizkin, B.A., Popovich, K., and Hartman, R. (2019). *Artificial neural network control of thermoelectrically-cooled microfluidics using computer vision based on IR thermography. Computers and Chemical Engineering* 121: 584–593.

103 Fudi, C., Hao, L., and Zhihan, X. (2015). *User-friendly optimization approach of fed-batch fermentation conditions for the production of iturin A using artificial neural networks and support vector machine. Electronic Journal of Biotechnology* 18 (4): 273–280.

104 Bouchareb, H., Semcheddine, S., Harmas, M. et al. (2019). *Virtual sensors to drive anaerobic digestion under a synergetic controller. Energies* 12 (3): 430.

105 Kravaris, C. and Savoglidis, G. (2012). *Tracking the singular arc of a continuous bioreactor using sliding mode control. Journal of the Franklin Institute* 349: 1583–1601.

106 Saeedizadeh, F., Pariz, N., and Hosseini, S.A. (2018). *High-order sliding mode control of a bioreactor model through non-commensurate fractional equations. Majlesi Journal of Electrical Engineering* 12 (1): 1–12.

107 Morales-Diaz, A., Vazquez-Sandoval, A.D., and Carlos-Hernandez, S. (2015). *Analysis and control of a distributed parameter reactor for pyrolysis of wood. Revista Mexicana de Ingeniería Química* 14 (2): 543–552.

108 Pundir, A.S. and Singh, K. (2019). *Temperature control of real-time identified fixed bed reactor by adaptive sliding mode control equipped with Arduino in Matlab. Asia-Pacific Journal of Chemical Engineering* 14: e2297.

Part III

Software Sensors and Observer-Based Control Schemes
for Bioprocess

5

Dynamical Behavior of a 3-Dimensional Continuous Bioreactor

The dynamic behavior of a two-dimensional model of a continuous bioreactor will be studied in this chapter. The state variables for the system are restricted to substrate and biomass concentrations, where the specific growth rate is a smooth function of the substrate concentration, which can be an algebraic equation, usually an approximate representation of the real data, i.e. with a behavior as a function that is increasing, or implies, decreasing (like Monod, Levenspiel, Haldane, Haldane-Boulton, Haldane-Luong, Moser–Boulton, Teissier, etc.). The effect of the input (dilution rate) on the multiplicity of equilibrium is shown in the open and closed loop configuration. The objective of the analysis under the control of feedback allows the most suitable regions to be found and model where the best performance of the bioreactor operates. Finally, the closed-loop instability of the inner or uncontrolled dynamics of the bioreactor is simulated. An example of a cell-producing bioreactor illustrates the simulated results.

5.1 Introduction

Much emphasis has been placed on the control of continuous and fed-batch bioreactors because of their prevalence in bioproducts, however, if production of cell mass or product is to be optimized, then a continuous operation is desirable in the development of bioprocess engineering. Unforeseen disturbances in this bioprocess (continuous bioreactors principally) may result in a failure in the operation of the reactor, which requires a new start-up procedure. Particularly, bioprocesses are quite interesting because they have become widely used in industries with different purposes, such as [1–3]:

- Producing chemical compounds synthesized by microorganisms in fermentative processes [4].
- Cultivating biomass for utilization or extraction of metabolites [5].
- Degrading a pollutant in wastewater/solid treatment [6].

In this sense, the dynamical analysis in bio-reactive systems, associated with a wide range of biological processes, requires greater attention in order to understand its dynamical nature as well as to promote operating conditions to improve performance at the pilot-plant level for its possible application at industrial level [2]. In this area, there are three interesting challenges to model:

Control in Bioprocessing: Modeling, Estimation and the Use of Soft Sensors, First Edition.
Pablo Antonio López Pérez, Ricardo Aguilar López, and Ricardo Femat.
© 2020 John Wiley & Sons Ltd. Published 2020 by John Wiley & Sons Ltd.

- understanding dynamics
- controlling dynamics
- observable dynamics.

The procedure of design and control of industrial nonlinear systems provides a good target for the identification of stable operating modes. Nonetheless, the ease of the regulation of important states will depend upon dynamic features related to operating conditions, design, and the closed-loop relationships between them [7]. Hence, one of the most important characteristics that hasa to be analyzed when a reacting system is to be controlled is the stability of the zero dynamics, which means the stability of the inverse of the process model [8]. In what follows, a strategy to track dynamic features of the zero dynamics of control affine nonlinear systems is developed.

As is well known, it is common to find control affine schemes for industrial bio/chemical reactors, which means that vector of manipulated variables (u) appears linearly in the model of the system (mass and energy balances). This model can be written in vector form as (5.1).

$$\dot{x} = f(x) + G(x)u \tag{5.1}$$

Here, $x \in \mathfrak{R}^n$ is the vector of states; $u \in \mathfrak{R}^q$; $f(x): \mathfrak{R}^n \to \mathfrak{R}^n$ is a nonlinear, smooth vector field; and $G(x): \mathfrak{R}^{n \times q}$ is a matrix that contains linear relationships between control and manipulates variables.

Now, in order to analyze the zero dynamics of a system, it is necessary to find the dynamics of its inverse [8, 9]. For nonlinear systems, the realization of this inverse process is often impossible. However, for control-affine systems that are partially controlled, it is possible to assess stability features of the zero dynamics when the system is regulated by the control of a subset of q states x_C, following the dynamics of the $(n - q)$ uncontrolled states x_D (Eq. (5.2)).

$$\dot{x} = f(x) + G(x)u \rightarrow \begin{cases} \dot{x}_C = f_C(x) + G_C(x)u \\ \dot{x}_D = f_D(x) + G_D(x)u \end{cases} \tag{5.2}$$

Here,

$$x_C \in \mathfrak{R}^q, \quad f_C(x) \equiv \{f_C : \mathfrak{R}^q \to \mathfrak{R}^q\}, \quad G_C(x) \equiv \{G_C : \mathfrak{R}^{q \times q}\}$$

$$x_D \in \mathfrak{R}^{n-q}, \quad f_D(x) \equiv \{f_D : \mathfrak{R}^{n-q} \to \mathfrak{R}^{n-q}\}, \quad G_D(x) \equiv \{G_D : \mathfrak{R}^{(n \times q)}\}$$

Now it is necessary to find theoretical values of manipulated inputs, u, assuming that the regulated variables will remain steady at the desired set point as a consequence of control action (Eq. (5.3)) If the evolution of the dynamic behavior of the uncontrolled variables (Eq. (5.4)) is not stable, it is possible to conclude that the zero dynamics are also not stable [10]; hence the control will exhibit poor performance during closed loop operation

$$\dot{x}_C^{sp}(t) = 0 \quad \Leftrightarrow \quad u^{sp} = -G_C^{-1}(x(t))f_C(x(t)) \tag{5.3}$$

$$\Rightarrow \dot{x}_D(t) = f_D(x(t)) - G_D(x(t))G_C^{-1}(x(t))f_C(x(t)) \tag{5.4}$$

Therefore, all the balances for the uncontrolled variables should tend to an attractor in order to ensure complete stability of the zero dynamics and, of course, of the control at the desired set point. The policy proposed is to ask for all the balances $\dot{x}_D(t)$ to exhibit negative sign (decreasing trend) when operating at the set point [2].

Proposition 5.1 *The Controller of the System $\dot{x}$ Will Be Stable $\forall t > 0$, If and Only If $\dot{x}_D < 0; \forall x_D \in x_D$*

Remarks

- The analysis of systems having more inputs than outputs, of systems in which normal forms cannot be defined, and of systems in which the zero dynamics are unstable is still a substantially unexplored and open to research [11].
- The concept of zero dynamics: a dynamical system that characterizes the internal behavior of a system once initial conditions and inputs are chosen in such a way as to constrain the output to be identically zero [12, 13].
- Furthermore, it should stressed that while in a linear system the asymptotic stability of the zero-dynamics proposes the property of input-to-state stability for the dynamics of the inverse system, with y viewed as input, this consequence is no longer true for a nonlinear system [12, 13].
- An invertible nonlinear system whose inverse is "state-free" can be turned into a fully linear controllable and observable system by means of dynamic state-feedback and coordinate transformations. Namely, if the value of the input $\forall t > 0$ can be expressed as a function of the values $t > 0$ of the output and a finite number of its derivatives [14, 15].

5.2 Bioreactor Modeling

From the bio-product systems point of view, the sulfate-reducing bacteria (SRB) are of great importance in environmental and industrial processes, which demand greater knowledge of the kinetic behavior of sulfate-reducing processes. However, the determination of kinetic parameters throughout the *structured model* (intracellular concentrations) on the basis of biomass components, such as the concentration of metabolites, enzymes, DNA, and/or RNA is a complex task [16–18]. For this reason, the kinetic parameters more commonly used are estimated through an *unstructured kinetic model* that uses measurable biomass, substrate, product, as well as yield coefficients determined in the bulk of the reactor [19, 20]. Few kinetic models have obtained satisfactory fitting of sulfate-reducing kinetic [21, 22], in most cases the Monod model is used, which doesn't take into account the product inhibition phenomenon generated by sulfide accumulation inside the bioreactor; much less the EPS production present at a later stage of the reacting paths.

5.2.1 Estimation of the Kinetic Parameters

The growth kinetic parameters were estimated by the rate of change of biomass production, using central finite differences according to the following equation:

$$\left(\frac{dt}{dt}\right)_{ti} \cong \left(\frac{\Delta X}{\Delta t}\right) = \left(\frac{X_{i+1} - X_i}{t_{i+1} - t_i}\right) \tag{5.5}$$

and a nonlinear multivariable regression for the rates of change of biomass production and experimental data (X, S, and P) was done. POLYMATH 6.0 Professional software was used, the program allows the application of effective numerical analysis techniques, and the Levenberg–Marquardt (L–M) algorithm was used for this case. Table 5.2 contains the parameter set obtained from the above methodology and Table 5.3 contains the structure of the kinetic rates and coefficients yield.

The L–M algorithm is one of the most widely used methods for solving nonlinear least squares problems. The kinetic model proposed in the bioprocess has been traditionally fit using graphical-based techniques, in which parameter estimation is converted to a linear regression problem. However, the applicability of this approach is related to the type of model as well as related to the complexity of the algebraic structure. Nonlinear regression was used based on the Levenberg–Marquardt least squares minimization algorithm, which is a hybrid of the Gauss–Newton and the steepest descent methods.

Biomass (X) mass balance:

$$\frac{dX}{dt} - DX + r_X - r_d \tag{5.6}$$

Sulfate (S) mass balance:

$$\frac{ds}{dt} = D(S_{in} - S) - r_X(Y_{S/X}) \tag{5.7}$$

Sulfide (P) mass balance:

$$\frac{dp}{dt} = -Dp + r_X(Y_{P/X}) \tag{5.8}$$

Where the corresponding kinetic rates and kinetic models are given in Tables 5.1 and 5.2.

Here $\mu(t)$ is the specific growth rate, which is a function of the concentrations; COD, biomass, acetate, and metal among others. Such reaction rates vary with time and are usually influenced by many physicochemical and biological environmental factors like substrate, biomass, and product concentrations as well as pH, temperature, dissolved oxygen concentration or various microbial growth inhibitors. Actually, there exists no systematic rule in order to determine the best model. Moreover, the identification of their biological parameters can be time-consuming. Indeed, these structures are nonlinear and, in most cases, nonlinearizable.

All these models describe the relationship between bacterial specific growth rate (μ) and substrate concentration. In other cases, combinations of two models were

Table 5.1 Kinetic rates and coefficient yields definitions.

Balance	Equation
Growth rate	$r_X = \mu X$
Dead rate	$r_d = \mu_d X$
Substrate coefficient yield	$Y_{S/X} = \dfrac{S_0 - S_i}{X_i - X_0}$
Product coefficient yield	$Y_{P/X} = \dfrac{P_i - P_0}{X_i - X_0}$

Table 5.2 Kinetic models for substrate and product inhibition.

Kinetic model	Equation	References
Haldane–Boulton	$\mu = \left[\dfrac{\mu_{max} S}{k_s + S + \left(S^2/k_i \right)} \right] \left[\dfrac{k_p}{k_p + P} \right]$	[23]
Moser–Luong	$\mu = \left[\dfrac{\mu_{max} S^n}{k_s + S^n} \right] \left[1 - \left(\dfrac{1}{k_p} \right)^m \right]$	[23]
Moser–Boulton	$\mu = \left[\dfrac{\mu_{max} S^n}{k_s + S^n} \right] \left[\dfrac{k_p}{k_p + P} \right]$	[23]
Levenspiel	$\mu = \mu_{max} \left[\dfrac{S}{k_s + S} \right] \left[1 - \dfrac{P}{P^*} \right]^n$	[23]

used: Haldane–Boulton, Haldane–Levenspiel, Haldane–Luong, Moser–Boulton, Moser–Levespiel, and Moser–Luong. Moosa et al., (2002) reported the effects of sulfate concentration and temperature on bacterial growth rate. Their experimental data were fitted with different mathematical models including those of Monod, Chen and Hashimoto, and Contoins [21, 24].

5.3 Main Results

In this section, the corresponding results of the proposed control method are shown. SRBs have a high impact on the ecological carbon and sulfur cycles. SRBs are widely distributed, and in environments with low sulfate levels, such as bodies of fresh water, they have relevance in the mineralization of organic matter.

An anaerobic bioreactor was simulated to evaluate the advantages of the proposed control law comparing the open-loop and the closed-loop performance. Numerical simulations were done employing the 23s MatlabTM R2009 library to solve ordinary differential equations, the following initial conditions for the corresponding concentrations were considered: $X_0 = 100 \, mg \, l^{-1}$, $S_0 = 5250 \, mg \, l^{-1}$, $P_0 = 20 \, mg \, l^{-1}$.

The experimental data served to obtain the nine parameters value of the model developed (see Table 5.3). The parameter values were determined by nonlinear regression employing the L–M (see Table 5.4). The L–M gradient descent method converges well for problems (least squares minimization algorithm), which is a hybrid of the Gauss–Newton and the steepest descent methods.

These statistics used absolute values rather than squared differences (as in their originally specified counterparts). Interpretation of correlation-based measures 0.95 indicates that the model explains 95.0% of the variability in the observed data.

Figures 5.1–5.3 show the dynamics of the sulfide, biomass, and sulfate concentrations, when the system represented by Eqs. (5.6)–(5.8) is under open-loop and closed-loop operation for a time interval of [0–75] hours, activating the control loop after 25 hours, in addition, the responses of the uncontrolled variables are shown in Figures 5.1 and 5.2.

Table 5.3 Kinetic parameters estimated for *Desulfovibrio alaskensis* 6SR.

Parameters	Value	Definition	Units
Haldane–Boulton			
μ_{max}	39.84 ± 10.01	Maximum specific rate	h^{-1}
K_S	86070 ± 100	Saturation constant for the Haldane–Boulton model	$mg\,l^{-1}$
K_i	9850 ± 19	Inhibition term for the Haldane–Boulton model	$mg\,l^{-1}$
K_P	7.24 ± 1	Inhibition term for the Haldane–Boulton model for the product	$mg\,l^{-1}$
Haldane–Luong			
μ_{max}	7 ± 1	Maximum specific rate	h^{-1}
K_S	2227 ± 10	Saturation constant for the Haldane–Luong model	$mg\,l^{-1}$
K_i	565 ± 20	Inhibition term for the Haldane–Luong model	$mg\,l^{-1}$
K_P	557 ± 30	Inhibition term for the Haldane–Luong model for the product	$mg\,l^{-1}$
m	0.10 ± 0.05	The exponential term for the Haldane–Luong model	Dimensionless
Moser–Boulton			
μ_{max}	10 ± 3	Maximum specific rate	h^{-1}
K_S	$1.26E+09 \pm 10$	Saturation constant for the Moser–Boulton model	$mg\,l^{-1}$
K_P	2.7 ± 0.1	Inhibition term for the Moser–Boulton model for the product	$mg\,l^{-1}$
n	2.53 ± 0.25	The exponential term for the Moser–Boulton model	Dimensionless
Levenspiel			
μ_{max}	0.36 ± 0.1	Maximum specific rate	h^{-1}
K_S	6550 ± 100	Saturation constant for the Moser–Boulton model	$mg\,l^{-1}$
P	610 ± 10	Inhibition concentration for the Moser–Boulton model for the product	$mg\,l^{-1}$
n	0.89 ± 0.20	The exponential term for the Moser–Boulton model	Dimensionless

Figure 5.3 shows the behavior of the four models proposed in closed loop, controlling the concentration of sulfate, which is verified by a disturbance in the setpoint (sp) in the form of two-step jumps $t > 25$ hours and $t > 50$ hours, and sp: $3200\,mg\,l^{-1}$ and sp: $3000\,mg\,l^{-1}$. The controller was linear proportional and the gain was chosen heuristically due to the nonlinear behavior of the k_p system: $1.5\,h^{-1}$.

We observe a smooth response for the Haldane–Bulton and Levenspiel model for the variables to be controlled, on the other hand, for the internal dynamics of the variables that are not controlled, only the Moser–Luong model showed a smooth and regulated response

Table 5.4 Correlation coefficients R^2.

Variable	Haldane–Boulton	Moser–Luong	Moser–Boulton	Levenspiel
X	0.98	0.97	0.95	0.94
S	0.97	0.96	0.96	0.95
P	0.97	0.97	0.96	0.96
Average	0.97	0.96	0.95	0.95

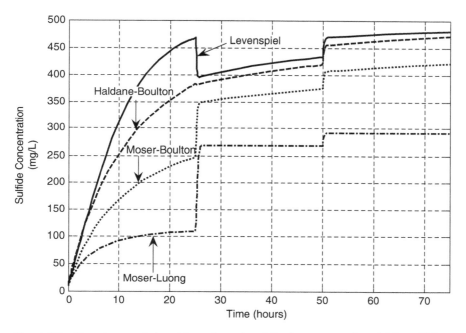

Figure 5.1 Open-loop and closed-loop dynamics of the four proposed model for sulfide concentrations.

while that for the other closed-loop models and under the same conditions of disturbance have a growing complexity (see, Figures 5.3 and 5.4).

Finally, a diagram in three dimensions is shown in Figure 5.6 which allows visualization of the different equilibrium points of the system, with the initial conditions but with a different scientific model, which allows the effect of the kinetic model as well as its closed loop effect in the non-controlled variables to be demonstrated.

Finally, considering the Haldane–Bulton model, it is necessary to know the behavior (reactive values – zero dynamics) of the uncontrolled closed-loop variables based on the conclusions of the introduction to the chapter. In the design of controllers, the relative degree and the stability of the zero dynamics of the control plant are usually assumed to be known in advance, and are determined by analyzing the system dynamic equations. Figure 5.7 shows the result of zero dynamics; these dynamics of nonlinear systems indicate

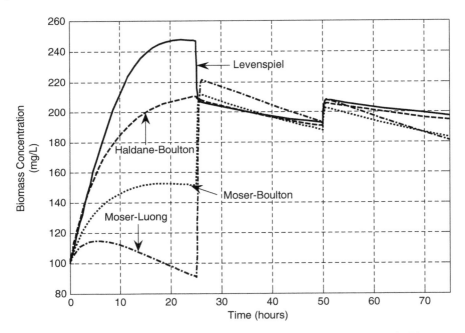

Figure 5.2 Open-loop and closed-loop dynamics of the four proposed model for biomass concentrations.

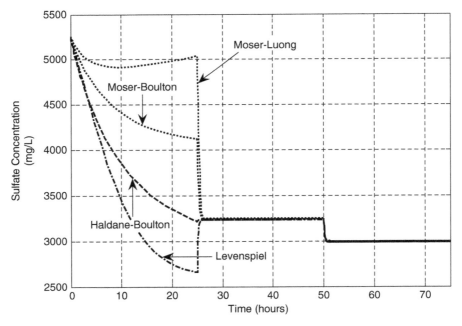

Figure 5.3 Open-loop and closed-loop dynamics of the four proposed model for sulfate concentrations.

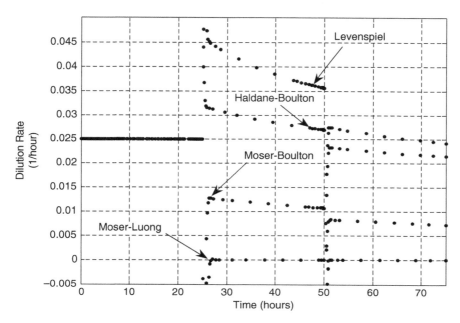

Figure 5.4 Open-loop and closed-loop dynamics of the four proposed model vs dilution rate.

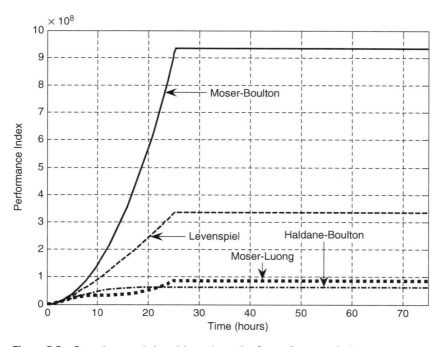

Figure 5.5 Open-loop and closed-loop dynamics for performance index.

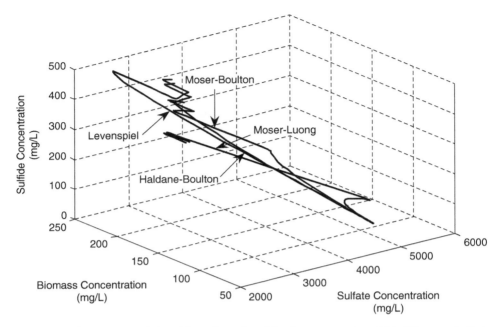

Figure 5.6 Open-loop and closed-loop dynamics of the four proposed model in 3D concentrations.

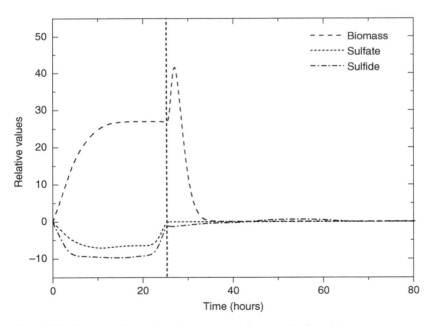

Figure 5.7 The zero dynamics performance of the uncontrolled states.

Table 5.5 Variability of the internal dynamics of each model with respect to the effort of control and performance index (see, figure 5).

Model	Effort of control	Performance Index X10^9
Haldane–Boulton	Acceptable	9.3
Moser–Luong	No acceptable	0.9
Moser–Boulton	Acceptable	0.7
Levenspiel	> regulation time and over peak	3.4

the stability properties of the closed system when the output is forced to be zero [25–28]. Thus, it is essential to assure that uncontrolled states reach a stable or unstable behavior. It should be noted that the time-derivatives converge to zero for sulfide (see, Table 5.5).

5.4 Concluding Remarks

The dynamic behavior for a class of continuous bioreactors in open-loop and closed loop-configuration are analyzed. The multiplicity of the equilibria are studied for growth rate kinetics (e.g. Monod, Teissier or Haldane models). It is necessary to design indicators that show which model has greater open-loop prediction ability as a closed loop, it is clear that the control effort in this work allows to give physical guidelines that model predicted real values for a possible operation in real time of the bioprocess.

List of Figures

List of Tables

References

1 Lara-Cisneros, G., Femat, R., and Perez, E. (2012). On dynamical behaviour of two-dimensional biological reactors. *International Journal of Systems Science* 43 (3): 526–534.

2 López-Pérez, P.A., Cuervo-Parra, J.A., Pérez-España, V.H. et al. (2018). Development of a novel kinetic model for cocoa fermentation applying the evolutionary optimization approach. *International Journal of Food Engineering* 14 (5-6).

3 Smith, H.L. and Waltman, P. (1995). *The Theory of the Chemostat: Dynamics of Microbial Competition*. Cambridge: Cambridge University Press.

4 Aguilar-López, R., López-Pérez, P.A., and Cuevas-Ortíz, F. (2011). *Esquemas de Control para Procesos Industriales: Diseño e implementación a Bio-procesos*. Editorial Académica Española.

5 Vuppaladadiyam, A.K., Prinsen, P., Raheem, A. et al. (2018). Microalgae cultivation and metabolites production: a comprehensive review. *Biofuels, Bioproducts and Biorefining* 12: 304–324.

6 Hernández-Melchor, D.J., López-Pérez, P.A., Carrillo-Vargas, S. et al. (2017). Experimental and kinetic study for lead removal via photosynthetic consortia using genetic algorithms to parameter estimation. *Environmental Science and Pollution Research* 25 (1).

7 Hermann, R. and Krener, A.J. (1977). Non-linear controllability and observability. *IEEE Transactions on Automatic Control* AC-22: 728–740.

8 Isidori, A. (2013). The zero dynamics of an online ar system: from the origin to the latest progresses of a long successful story. *European Journal of Control* 19: 369–378.

9 Neria-González, M.I., López-Pérez, P.A., and Aguilar-López, R. (2016). Partial control of a continuous bioreactor: application to an anaerobic system for heavy metal removal. *Journal of Engineering* 2016.

10 Isidori, A. (2011), *The zero dynamics of a nonlinear system: from the origin to the latest progresses of a long successful story*, 30th Chinese Control Conference 23 July 2011.

11 Byrnes, L.C.I. and Isidori, A. (1984). A frequency domain philosophy for nonlinear systems. *IEEE Conference on Decision and Control* 23: 1569–1573.

12 Isidori, A. (2013). The zero dynamics of a nonlinear system: from the origin to the latest progresses of a long successful story. *European Journal of Control* 19: 369–378.

13 Liberzon, D., Morse, A.S., and Sontag, E.D. (2002). Output-input stability and minimumphase nonlinear systems. *IEEE Transactions on Automatic Control* AC-43: 422–436.

14 Isidori, A., Moog, C. H. and Luca, A. D. (1986), *A sufficient condition for full linearization via dynamic state feedback*, 1986 25th IEEE Conference on Decision and Control, Athens, Greece, 203–208.

15 Byrnes, C. I., Isidori, A. (1984), *A frequency domain philosophy for nonlinear systems with applications to stabilization and adaptive control*, January 1985, Source IEEE Xplore, Conference: Decision and Control, The 23rd IEEE Conference on Volume: 23.

16 Grijalva-Hernández, F., Peña-Caballero, V., López-Pérez, P.A., and Aguilar-López, R. (2018). Estimation of plasmid concentration in batch culture of *Escherichia coli* DH5α via simple state observer. *Chemical Papers* 72 (10): 2589–2598.

17 Pentjuss, A., Stalidzans, E., Liepins, J. et al. (2017). Model-based biotechnological potential analysis of *Kluyveromyces marxianus* central metabolism. *Journal of Industrial Microbiology & Biotechnology* 44: 1177–1190.

18 Almquist, J., Cvijovic, M., Hatzimanikatis, V. et al. (2014). Kinetic models in industrial biotechnology — improving cell factory performanc. *Metabolic Engineering* 24: 38–60.

19 Gamboa-Rueda, J.A., Lizcano-González, V.A., Ordoñez-Supelano, M.A. et al. (2015). Unstructured kinetic model for batch fermentation of USP glycerol for lactic acid production. *CT&F - Ciencia, Tecnología y Futuro* 6 (1): 81–94.

20 Kasbawati, S.S., Jaya, A.K., and Kalondeng, A. (2018). Determining an appropriate unstructured kinetic model for batch ethanol fermentation data using a direct search method. *Biotechnology & Biotechnological Equipment* https://doi.org/10.1080/13102818.2018.1503563.

21 Moosa, S., Nemati, M., and Harrison, S.T.L. (2002). A kinetic study on anaerobic reduction of sulphate, Part I: Effect of sulphate concentration. *Chemical Engineering Science* 57 (14): 2773–2780.

22 Velázquez-Sánchez, H. I., López-Pérez, P. A., Neria-González, M. I. and Aguilar-López, R. (2016), *Enhancement of Bio-Hydrogen Production Technologies by Sulphate-Reducing Bacteria. Hydrogen Generation Technologies*, Enhacement of Biohydrogen Production Technologies by Sulphate-Reducing Bacteria.; WILEY-Scrivener.

23 Neeleman, R. (2002), *Biomass performance: Monitoring and control in Bio-pharmaceutical production*, Ph. D Thesis, Wageningen University. 15–16.

24 Moosa, S., Nemati, M., and Harrison, S.T.L. (2005). A kinetic study on anaerobic reduction of sulphate, Part II, Incorporation of temperature effects in the kinetic model. *Chemical Engineering Science* 60: 3517–3524.

25 Nguyen, D.H. (2018). Minimum-rank dynamic output consensus design for heterogeneous nonlinear multi-agent systems. *IEEE Transactions on Control of Network Systems* 5: 105–115.

26 Wu, W. (2000). Nonlinear bounded control of a nonisothermal CSTR. *Industrial & Engineering Chemistry Research* 39: 3789–3798.

27 Galluzzo, M. and Cosenza, B. (2011). Control of a non-isothermal continuous stirred tank reactor by a feedback–feedforward structure using type-2 fuzzy logic controllers. *Information Sciences* 181: 3535–3550.

28 Ho, Y.K., Mjalli, F.S., and Yeoh, H.K. (2010). Multivariable adaptive predictive model based control of a biodiesel transesterification reactor. *Journal of Applied Sciences* 10: 1019–1027.

6

Observability Analysis Applied to 2D and 3D Bioreactors with Inhibitory and Non-inhibitory Kinetics Models

The bioreactor mathematical models are employed to describe and predict the dynamics of its key state variables, such as metabolites, substrates, and biomass concentrations. These models are also used in the design, optimization, on-line monitoring, and control of bioprocesses. Observability analysis applied to 2D and 3D bioreactors with inhibitory and non-inhibitory models is studied in this chapter. The chemostat in 2D (batch or continuous) will always be observable if even one of the observability sub-matrix determinants is different from zero, with the only condition that the growth rate is non-zero. The chemostat in 3D is non-observable no matter whether the reactor is operated in a batch or continuous bioreactor. Sufficient and necessary conditions for state observability of such system was established. These results indicated that the proposed model can be applied for simulations on different conditions of operation for possible instrumentation, estimation, and control from laboratory scale up to semi-pilot scale.

6.1 Introduction

The bioreactor mathematical models are employed to describe and predict the dynamics of its key state variables, such as metabolites, substrates, and biomass concentrations. These models are also used in the design, optimization, on-line monitoring, and control of bioprocesses. To calculate the global rate for some biochemical reactions that together transform at least one substrate to biomass and metabolites, mass and energy balances have been formulated where the global rate is modeled frequently with logistic-type mathematical functions, known as unstructured growth models [1]. In this way, the chemostat is the simplest bioreactor model that describes a microorganism culture [2], where a substrate is fed continuously into the bioreactor, which is consumed by the biomass and it is drawn off with the same input velocity. A minimum of two key states are regarded in a chemostat mass balance; the biomass and substrate concentrations [3]. In spite of this relative simplicity, the chemostat is very useful in many biological and applied mathematical studies [4].

State observation is certainly a fundamental problem in bioprocess monitoring and control, even more in distributed parameter bioreactors. As well as the question of designing efficient state estimators, the information contents of on-line measurements is crucial. The observer design problem is to analyze the observability conditions of the nonlinear systems. For linear systems, the classical observability index as the observability matrix for

Control in Bioprocessing: Modeling, Estimation and the Use of Soft Sensors, First Edition.
Pablo Antonio López Pérez, Ricardo Aguilar López, and Ricardo Femat.
© 2020 John Wiley & Sons Ltd. Published 2020 by John Wiley & Sons Ltd.

observability analysis and the estimator design have been extensively studied, and have proven extremely useful, especially for on-line monitoring and control applications such as observer based control design [5]. The design of observability conditions for nonlinear systems is a challenging problem (even for accurately known systems) that has received a considerable amount of attention. The first category of techniques consists of applying linear algorithms to the system linearized around the estimated trajectory [6].

Furthermore, observability is a clearly critical issue in dynamic systems in general and in chemical and biochemical systems in particular (e.g. Morari and Stephanopoulos [7]). The test of a system's observability is a necessary prerequisite to the estimation of states. Because of the nonlinear aspects of their dynamics, stability, and observability analysis is rather complex in (bio-chemical) process applications. However, for nonlinear systems, the theory of observers is not nearly as complete or successful as it is for linear systems.

The test of a system's observability is a necessary prerequisite to the estimation and regulation of states. Because of the nonlinear aspects of their dynamics, stability, and observability analysis is rather complex in (bio-chemical) process applications.

From the above the stability and observability properties of cellulose hydrolysis model is locally analyzed. Several combinations of measured outputs are analyzed in order to show the corresponding observability conditions of the process. Observability analysis applied to 2D and 3D bioreactor with inhibitory and non-inhibitory models is studied in this section.

6.2 Materials and Methods

6.2.1 Kinetic Models of Inhibition

Bioethanol production from cellulose hydrolysis is a promising alternative energy source, only a small percentage of all the microorganisms around the earth can degrade cellulose, and they are mainly bacteria and fungi [8]. Only a handful of works relating to the mathematical modeling of cellulose hydrolysis by microorganisms are reported in literature, for example, Agarwal et al. [9] calculated the value of the kinetic parameters for a set of growth kinetic models; all of these describe the carboxymethyl-cellulose hydrolysis by *Cellulomonas cellulans* in a batch culture. These growth models and their parameter values were taken for the development of this work (Table 6.1).

The chemostat model studied here, considers the biomass yield as a constant value or as a function of the substrate concentration [10]. Some restrictions for the modeling are that it is bounded in the positive quadrant (mass concentrations cannot take negative values experimentally); is isothermal and homogeneous in the reactant concentrations in the entire vessel; no terms of death rate were considered and it is governed by the principle of mass conservation, PMC [11]. From a mass balance for the substrate and biomass in the bioreactor, the following system is obtained:

$$\frac{ds}{dt} = f(s,x) = DS_i - DS - \frac{\mu(S)X}{Y} \tag{6.1}$$

$$\frac{dX}{dt} = g(s,x) = -DX + \mu(S)X \tag{6.2}$$

Table 6.1 Kinetic models of inhibition.

Model	Equation	Parameters
Aiba	$\mu = \mu_{max} \dfrac{S}{(Ks + S)(1 + {}^S\!/_{Ki})}$	$\mu_{max} = 0.383$, $Ks = 2.40$, $Ki = 6.569$
Andrew	$\mu = \mu_{max} \dfrac{S}{(Ks + S)}\left(1 - \dfrac{S}{S*}\right)^n$	$\mu_{max} = 0.412$, $Ks = 2.40$, $Ki = 2.5$
Haldane	$\mu = \mu_{max} \dfrac{\left(1 - \dfrac{S}{Cs}\right)^n S}{S + Kl\left(1 - \dfrac{S}{Cs}\right)^m}$	$\mu_{max} = 27.52$, $Ks = 388.3$, $Ki = 0.0256$
Han–Levenspiel	$\mu = \mu_{max} \dfrac{S}{\left(Ks + S + {}^{S^2}\!/_{Ki}\right)}$	$\mu_{max} = 0.182$, $S^2 = 388.5$, $n = 42.27$, $m = 1202$, $Km = 544.6$
Luong	$\mu = \mu_{max} \dfrac{S^n}{Ks + S^n}$	$\mu_{max} = 0.385$, $Ks = 3.7$, $S^n = 103736$, $n = 15718$
Moser	$\mu = \mu_{max} \dfrac{S}{Ks + S} e^{-S/Ki}$	$\mu_{max} = 1.108$, $Ks = 0.0015$, $n = -2.66$

6.2.1.1 Dynamics Models
Batch **2D**

$$\frac{dx}{dt} = \mu x \tag{6.3}$$

$$\frac{ds}{dt} = -\frac{\mu x}{Y} \tag{6.4}$$

3D

$$\frac{dx}{dt} = mux \tag{6.5}$$

$$\frac{ds}{dt} = -\frac{mux}{Y} \tag{6.6}$$

$$\frac{dp}{dt} = -\frac{mux}{Y_p} \tag{6.7}$$

Continuous Models **2D**

$$\frac{dx}{dt} = mux - Dx \tag{6.8}$$

$$\frac{ds}{dt} = D(S_o - s) - \frac{mux}{Y} \tag{6.9}$$

3D

$$\frac{dx}{dt} = mux - Dx \tag{6.10}$$

$$\frac{ds}{dt} = D(S_o - s) - \frac{mux}{Y} \tag{6.11}$$

$$\frac{dp}{dt} = \frac{mux}{Y_p} - Dp \tag{6.12}$$

Where S_i is the feed substrate concentration, in this case carboxymethyl-cellulose (CMC), (Kg m^{-3}); S is the substrate concentration in the reaction mixture (Kg m^{-3}); X is the biomass concentration (Kg m^{-3}). In this contribution the specific growth rate μ, (h^{-1}) is a function $\mu\colon [0,S_{max}] \to R$ with the following properties:

- μ is a differentiable function in the domain $[0,S_{max}]$.
- $\mu(0) = 0$, and $\mu(S) \leq \overline{\mu}_{max}$, where $\overline{\mu}_{max}$ is a scalar providing the upper bound of μ; D is the dilution rate (h^{-1}); Y is the biomass yield, $(\text{Kg}_{\text{biomass}}\ \text{Kg}_{\text{CMC}}^{-1})$. $S, X, D, Y \in \mathfrak{R}^+$.
- Biologically the initial conditions for the biomass and substrate concentrations at each time are: $X_0(t)$, $S_0(t) \geq 0$; $t \in [0, \infty)$.

6.2.2 Observability Criterion

The observability calculation for a linear and time-invariant (LTI) system depends on matrices A and C of the system. A LTI system is represented for the dynamic state equation and the output vector:

$$\dot{x}(t) = Ax(t) + Bu(t) \tag{6.13}$$

$$y(t) = Cx(t) \tag{6.14}$$

where $A \in \mathfrak{R}^{nxn}$, $C \in \mathfrak{R}^{rxn}$, is observable if and only if the observability matrix (Ob) has rank n:

$$rank(Ob) = rank \left(\begin{bmatrix} C \\ \cdots \\ CA \\ \cdots \\ \vdots \\ \cdots \\ CA^{n-1} \end{bmatrix} \right) = n \tag{6.15}$$

Observability indicates the ability to estimate the historical values of a state from the knowledge of the input and output variables of the system. The observability is critical for those systems where is impossible to know the complete state vector so it requires the estimation of this from output variables (see, Chapter 3).

A state x_i is observable in t_0 always so that it possible calculate $x_i(t_0)$ knowing $y(t)$.

If this property is satisfied $\forall\, t$ and $\forall\, i = 1,\ldots,n$ then the system is fully observable.

6.3 Results and Discussion

Both the observability analysis and the reconstruction or estimation of the non-measurable variables has been realized on linearized bioreactor kinetic equations, where the linearization is made around a specific equilibrium point.

Study Case 6.1 *Observability Matrix for Chemostat Model in Continuous Operation*

$$A = \begin{bmatrix} -D - \dfrac{\mu(s)'x}{Y}, & -\dfrac{\mu(s)}{Y} \\ \mu(s)'x & \mu(s) - D \end{bmatrix}; \quad C = \begin{bmatrix} 1 \\ 0 \end{bmatrix}_T ; \tag{6.16}$$

rank(A)=2 ∴The system is invertible

The largest array dimension for matrix A is two, hence the rank of the observability matrix (Ob) will be two in order for the system to be observable.

$$Ob = \begin{bmatrix} 1, & 0 \\ -D - \dfrac{\mu(s)'x}{Y}, & -\dfrac{\mu(s)}{Y} \\ \left(D + \dfrac{\mu(s)'x}{Y}\right)^2 - \dfrac{\mu(s)\mu(s)'x}{Y}, & \dfrac{\mu(s)\left(D + \frac{\mu(s)'x}{Y}\right)}{Y} + \dfrac{\mu(s)(D - \mu(s))}{Y} \end{bmatrix}; \tag{6.17}$$

rank(Ob)=2 ∴The system is observable

If $\mu(s) = D$ in the stationary state so,

$$A = \begin{bmatrix} -D - \dfrac{\mu(s)'x}{Y}, & -\dfrac{\mu(s)}{Y} \\ \mu(s)'x & 0 \end{bmatrix}; \quad C = \begin{bmatrix} 1 \\ 0 \end{bmatrix}_T ; \tag{6.18}$$

rank(A)=2 ∴The system is invertible

$$Ob = \begin{bmatrix} 1, & 0 \\ -D - \dfrac{\mu(s)'x}{Y}, & -\dfrac{\mu(s)}{Y} \\ \left(D + \dfrac{\mu(s)'x}{Y}\right)^2 - \dfrac{\mu(s)\mu(s)'x}{Y}, & \dfrac{\mu(s)\left(D + \frac{\mu(s)'x}{Y}\right)}{Y} \end{bmatrix}; \tag{6.19}$$

rank(Ob)=2 ∴The system is observable

Matrix Rank by Determinant

The observability matrix will have rank 2 if there is any square submatrix of order 2, such that its determinant is not zero.

$$Det_1 = \begin{bmatrix} 1, & 0 \\ -D - \dfrac{\mu(s)'x}{Y}, & -\dfrac{\mu(s)}{Y} \end{bmatrix} = -\dfrac{\mu(s)}{Y} \tag{6.20}$$

$$Det_2 = \begin{bmatrix} 1, & 0 \\ \left(-D - \dfrac{\mu(s)'x}{Y}\right)^2 - \dfrac{\mu(s)\mu(s)'x}{Y}, & \dfrac{\mu(s)\left(D + \frac{\mu(s)'x}{Y}\right)}{Y} \end{bmatrix} = -\dfrac{\mu(s)(\mu(s)'x + dY)}{Y^2}$$

$$Det_3 = \begin{bmatrix} -D - \dfrac{\mu(s)'x}{Y}, & -\dfrac{\mu(s)}{Y} \\ \left(D + \frac{\mu(s)'x}{Y}\right)^2 - \dfrac{\mu(s)\mu(s)'x}{Y}, & \dfrac{\mu(s)\left(D + \frac{\mu(s)'x}{Y}\right)}{Y} \end{bmatrix} = -\dfrac{\mu(s)(\mu(s)'x)}{Y^2} \tag{6.21}$$

For the above, the only way for the chemostat system (in continuous operation) to lose observability is that all the determinants of the submatrices are null, this case occurs when $\mu(s) = 0$, which means that there is no growth of microorganisms.

Study Case 6.2 *Observability Matrix for Chemostat Model in Continuous Operation, Considering Cell Death*

$$A = \begin{bmatrix} -D - \dfrac{\mu(s)'x}{Y}, & -\dfrac{\mu(s)}{Y} \\[2mm] \mu(s)'x & \mu(s) - D - kd \end{bmatrix}; \quad C = \begin{bmatrix} 1 \\ 0 \end{bmatrix}_T ; \tag{6.22}$$

rank(A)=2 ∴The system is invertible

$$Ob = \begin{bmatrix} 1, & 0 \\[2mm] -D - \dfrac{\mu(s)'x}{Y}, & -\dfrac{\mu(s)}{Y} \\[3mm] \left(D + \dfrac{\mu(s)'x}{Y}\right)^2 - \dfrac{\mu(s)\mu(s)'x}{Y}, & \dfrac{\mu(s)\left(D + \frac{\mu(s)'x}{Y}\right)}{Y} + \dfrac{\mu(s)(D + Kd - \mu(s))}{Y} \end{bmatrix}; \tag{6.23}$$

rank(Ob)=2 ∴The system is observable
  If $\mu(s) = D$ in the stationary state so,

$$A = \begin{bmatrix} -D - \dfrac{\mu(s)'x}{Y}, & -\dfrac{\mu(s)}{Y} \\[2mm] \mu(s)'x & -kd \end{bmatrix}; \quad C = \begin{bmatrix} 1 \\ 0 \end{bmatrix}_T ; \tag{6.24}$$

rank(A)=2 ∴The system is invertible

$$Ob = \begin{bmatrix} 1, & 0 \\[2mm] -D - \dfrac{\mu(s)'x}{Y}, & -\dfrac{\mu(s)}{Y} \\[3mm] \left(D + \dfrac{\mu(s)'x}{Y}\right)^2 - \dfrac{\mu(s)\mu(s)'x}{Y}, & \dfrac{\mu(s)\left(D + \frac{\mu(s)'x}{Y}\right)}{Y} + \dfrac{\mu(s)(Kd)}{Y} \end{bmatrix}; \tag{6.25}$$

rank(Ob)=2 ∴The system is observable

Observability Matrix for Chemostat Model in Batch Operation

$$A = \begin{bmatrix} -D - \dfrac{\mu(s)'x}{Y}, & -\dfrac{\mu(s)}{Y} \\[2mm] \mu(s)'x & \mu(s) \end{bmatrix}; \quad C = \begin{bmatrix} 1 \\ 0 \end{bmatrix}_T ; \tag{6.26}$$

rank(A)=1 ∴The system is not invertible; Det (A)=0

$$Ob = \begin{bmatrix} 1, & 0 \\[2mm] -\dfrac{\mu(s)'x}{Y}, & -\dfrac{\mu(s)}{Y} \\[3mm] \left(\dfrac{\mu(s)'x}{Y}\right)^2 - \dfrac{\mu(s)\mu(s)'x}{Y}, & \dfrac{\mu(s)\mu(s)'x}{Y^2} + \dfrac{\mu(s)^2}{Y} \end{bmatrix}; \tag{6.27}$$

rank(Ob)=2 ∴The system is observable

Matrix Rank by Determinant

The observability matrix will have rank 2 if there is any square submatrix of order 2, such that its determinant is not zero.

$$Det_1 = \begin{bmatrix} 1, & 0 \\ -D - \dfrac{\mu(s)'x}{Y}, & -\dfrac{\mu(s)}{Y} \end{bmatrix} = -\dfrac{\mu(s)}{Y} \tag{6.28}$$

$$Det_2 = \begin{bmatrix} 1, & 0 \\ \left(\dfrac{\mu(s)'x}{Y}\right)^2 - \dfrac{\mu(s)\mu(s)'x}{Y}, & \dfrac{\mu(s)\mu(s)'x}{Y^2} - \dfrac{\mu(s)^2}{Y} \end{bmatrix} = -\dfrac{\mu(s)(Y\mu(s) - \mu(s)'x)}{Y^2}. $$

$$Det_3 = \begin{bmatrix} -\dfrac{\mu(s)'x}{Y}, & -\dfrac{\mu(s)}{Y} \\ \left(\dfrac{\mu(s)'x}{Y}\right)^2 - \dfrac{\mu(s)\mu(s)'x}{Y}, & \dfrac{\mu(s)\mu(s)'x}{Y^2} - \dfrac{\mu(s)^2}{Y} \end{bmatrix} = 0 \tag{6.29}$$

For the above, the only way for the chemostat system (in batch operation) to lose observability is that all the determinants of the submatrices are null, this case occurs when $\mu(s) = 0$, which means that there is no growth of microorganisms.

Study Case 6.3 Observability Matrix for Chemostat Model in Batch Operation, Considering Yield Not Constant

$$A = \begin{bmatrix} -\dfrac{x(Y\mu(s)' - Yd\mu(s))}{Y^2}, & -\dfrac{\mu(s)}{Y} \\ \mu(s)'x & \mu(s) \end{bmatrix}; \quad C = \begin{bmatrix} 1 \\ 0 \end{bmatrix}_T ; \tag{6.30}$$

rank(A)=2 ∴The system is invertible

$$Ob = \begin{bmatrix} 1, & 0 \\ -\dfrac{x(Y\mu(s)' - Yd\mu(s))}{Y^2}, & -\dfrac{\mu(s)}{Y} \\ \left(-\dfrac{x(Y\mu(s)' - Yd\mu(s))}{Y^2}\right)^2 - \dfrac{\mu(s)\mu(s)'x}{Y}, & \dfrac{x\mu(s)(Y\mu(s)'}{Y^3} - \dfrac{\mu(s)^2}{Y} \end{bmatrix}; \tag{6.31}$$

rank(Ob)=2 ∴The system is observable

Study Case 6.4 Observability Matrix for Chemostat Model in Batch Operation

$$A = \begin{bmatrix} -\dfrac{\mu(s)'}{Y}, & -\dfrac{\mu(s)}{Y} \\ \mu(s)'x & \mu(s) - kd \end{bmatrix}; \quad C = \begin{bmatrix} 1 \\ 0 \end{bmatrix}_T ; \tag{6.32}$$

rank(A)=2 ∴The system is invertible

$$Ob = \begin{bmatrix} 1, & 0 \\ -\dfrac{\mu(s)'x}{Y}, & -\dfrac{\mu(s)}{Y} \\ \left(-\dfrac{\mu(s)'x}{Y}\right)^2 - \dfrac{\mu(s)\mu(s)'x}{Y}, & \dfrac{\mu(s)\mu(s)'x}{Y^2} + \dfrac{\mu(s)(\mu(s) - kd)}{Y} \end{bmatrix}; \tag{6.33}$$

rank(Ob)=2 ∴The system is observable

Study Case 6.5 Observability Matrix for Chemostat Model in Continuous Operation Applied to a Set of Unstructured Kinetic Models Not Constant Yield Coefficient

Monod Kinetics

$$A = \left[\begin{array}{cc} \dfrac{mmax1\,s\,x}{Y(Ks1+s)^2} - \dfrac{mmax1\,x}{Y(Ks1+s)} - D, & -\dfrac{mmax1\,s}{Y(Ks1+s)} \\[3ex] \left(D + \dfrac{mmax1\,x}{Y(Ks1+s)} - \dfrac{mmax1\,s\,x}{Y(Ks1+s)^2}\right)^2 - \dfrac{mmax1\,s\left(\dfrac{mmax1\,x}{Ks1+s} - \dfrac{mmax1\,s\,x}{(Ks1+s)^2}\right)}{Y\,(Ks1+s)}, & mmax1\,s\left(\dfrac{D + \dfrac{mmax1\,x}{Y(Ks1+s)} - \dfrac{mmax1\,s\,x}{(Ks1+s)^2}}{Y\,(Ks1+s)}\right) \end{array} \right]$$

Aiba Kinetics

$$A = \left[\begin{array}{cc} \dfrac{mmax2\,s\,x}{Ye^{\frac{s}{Ki}}(Ks2+s)^2} - \dfrac{mmax2\,x}{Ye^{\frac{s}{Ki}}(Ks2+s)} - D + \dfrac{mmax2\,s\,x}{Y\,Ki\,e^{\frac{s}{Ki}}(Ks2+s)}, & -\dfrac{mmax2\,s}{Ye^{\frac{s}{Ki}}(Ks2+s)} \\[3ex] \left(D + \dfrac{mmax2\,x}{Ye^{\frac{s}{Ki}}(Ks2+s)} - \dfrac{mmax2\,s\,x}{Ye^{\frac{s}{Ki}}(Ks2+s)^2} - \dfrac{mmax2\,s\,x}{Ki\,Ye^{\frac{s}{Ki}}(Ks2+s)}\right)^2 & mmax2\,s\left(D + \dfrac{mmax2\,x}{Ye^{\frac{s}{Ki}}(Ks2+s)} - \dfrac{mmax2\,sx}{Ye^{\frac{s}{Ki}}(Ks2+s)^2} + \dfrac{mmax2\,sx}{Ki\,Ye^{\frac{s}{Ki}}(Ks2+s)}\right) \\ {} + \dfrac{mmax2\,s\left(\dfrac{mmax2\,s\,x}{e^{\frac{s}{Ki}}(Ks2+s)^2} - \dfrac{mmax2\,x}{e^{\frac{s}{Ki}}(Ks2+s)} + \dfrac{mmax2\,x}{Ki\,e^{\frac{s}{Ki}}(Ks2+s)}\right)}{Y\,e^{\frac{s}{Ki}}(Ks2+s)}, & \Big/\ \big(Y\,e^{\frac{s}{Ki}}(Ks2+s)\big) \end{array} \right]$$

Moser Kinetics

$$A = \left[\begin{array}{cc} \dfrac{mmax7\,s^{u}s^{u-1}u\,x}{Y(Ks6+s^u)^2} - \dfrac{mmax7\,s^{u-1}u\,x}{Y(Ks6+s^u)} - D, & -\dfrac{mmax7\,s^{u}}{Y(Ks6+s^u)} \\[3ex] \left(D + \dfrac{mmax7\,s^{u-1}u\,x}{Y(Ks6+s^u)} - \dfrac{mmax7\,s^{u}s^{u-1}u\,x}{Y(Ks6+s^u)^2}\right)^2 - \dfrac{mmax7\,s^{u}\left(\dfrac{mmax7\,s^{u-1}u\,x}{Ks6+s^u} - \dfrac{mmax7\,s^{u}s^{u-1}u\,x}{(Ks6+s^u)^2}\right)}{Y(Ks6+s^u)}, & mmax7\,s^{u}\left(\dfrac{D + \dfrac{mmax7\,s^{u-1}u\,x}{Y(Ks6+s^u)} - \dfrac{mmax7\,s^{u}s^{u-1}u\,x}{(Ks6+s^u)^2}}{Y(Ks6+s^u)}\right) \end{array} \right]$$

Haldane Kinetics

$$A = \left[\begin{array}{cc}
\dfrac{mmax4\,s\,x\left(\frac{2s}{Kih}+1\right)}{Y\left(Ks4+s+\frac{s^2}{Kih}\right)^2} - \dfrac{mmax4\,x}{Y\left(Ks4+s+\frac{s^2}{Kih}\right)} - D, & -\dfrac{mmax4\,x}{Y\left(Ks4+s+\frac{s^2}{Kih}\right)} \\[2em]
mmax4\,s\left(D + \dfrac{mmax4\,x}{Y\left(Ks4+s+\frac{s^2}{Kih}\right)} - \dfrac{mmax4\,s\,x\left(\frac{2s}{Kih}+1\right)}{Y\left(Ks4+s+\frac{s^2}{Kih}\right)^2}\right), & \dfrac{mmax4\,s\,x\left(\frac{2s}{Kih}+1\right)}{Y\left(Ks4+s+\frac{s^2}{Kih}\right)} - mmax4\,s\,\dfrac{\frac{mmax4\,x}{Ks4+s+\frac{s^2}{Kih}}}{Y\left(Ks4+s+\frac{s^2}{Kih}\right)}
\end{array}\right],$$

Andrew Kinetics

$$A = \left[\begin{array}{cc}
\dfrac{mmax3\,s\,x}{Y\,(Ks3+s)^2\left(\frac{s}{Kin}+1\right)} - \dfrac{mmax3\,x}{Y\,(Ks3+s)\left(\frac{s}{Kin}+1\right)} - D + \dfrac{mmax3\,s\,x}{Kin\,Y\,(Ks3+s)\left(\frac{s}{Kin}+1\right)^2}, & -\dfrac{mmax3\,s}{Y\,(Ks3+s)\left(\frac{s}{Kin}+1\right)} \\[2em]
mmax3\,s\left(D + \dfrac{mmax3\,x}{Y\,(Ks3+s)\left(\frac{s}{Kin}+1\right)} - \dfrac{mmax3\,s\,x}{Y\,(Ks3+s)^2\left(\frac{s}{Kin}+1\right)} - \dfrac{mmax3\,s\,x}{Kin\,Y\,(Ks3+s)\left(\frac{s}{Kin}+1\right)^2}\right) + \dfrac{\frac{mmax3\,s\,x}{(Ks3+s)\left(\frac{s}{Kin}+1\right)} - \frac{mmax3\,s\,x}{(Ks3+s)^2\left(\frac{s}{Kin}+1\right)} + \frac{mmax3\,s\,x}{Kin\,(Ks3+s)\left(\frac{s}{Kin}+1\right)^2}}{Y\,(Ks3+s)\left(\frac{s}{Kin}+1\right)}, & \dfrac{mmax3\,s}{Y\,(Ks3+s)\left(\frac{s}{Kin}+1\right)}
\end{array}\right],$$

Han–Levenspiel Kinetics

$$A = \begin{bmatrix} a & b \\ c & d \end{bmatrix}$$

$$a = -D - \frac{mmax5\,x\left(1 - \dfrac{s}{Sinh}\right)^{n}}{Y\left(s + Km\left(1 - \dfrac{s}{Sinh}\right)^{m}\right)} - mmax5s\,x\,\frac{\left(\dfrac{Kmm\left(1 - \dfrac{s}{Sinh}\right)^{m-1}}{Sinh} - 1\right)\left(1 - \dfrac{s}{Sinh}\right)^{n}}{Y\left(s + Km\left(1 - \dfrac{s}{Sinh}\right)^{m}\right)^{2}} + \frac{mmax5nsx\left(1 - \dfrac{s}{Sinh}\right)^{n-1}}{SinhY\left(s + Km\left(1 - \dfrac{s}{Sinh}\right)^{m}\right)}$$

$$b = D + \frac{mmax5\,x\left(1 - \dfrac{s}{Sinh}\right)^{n}}{Y\left(s + Km\left(1 - \dfrac{s}{Sinh}\right)^{m}\right)} + mmax5s\,x\,\frac{\left(\dfrac{Kmm\left(1 - \dfrac{s}{Sinh}\right)^{m-1}}{Sinh} - 1\right)\left(1 - \dfrac{s}{Sinh}\right)^{n}}{Y\left(s + Km\left(1 - \dfrac{s}{Sinh}\right)^{m}\right)^{2}} + \frac{mmax5nsx\left(1 - \dfrac{s}{Sinh}\right)^{n-1}}{SinhY\left(s + Km\left(1 - \dfrac{s}{Sinh}\right)^{m}\right)} - \cdots$$

$$\left\{ mmax5s\left(1 - \dfrac{s}{Sinh}\right)^{n}\left[\frac{mmax5\,x\left(1 - \dfrac{s}{Sinh}\right)^{n}}{s + Km\left(1 - \dfrac{s}{Sinh}\right)^{m}} + mmax5s\,x\,\frac{\left(\dfrac{Kmm\left(1 - \dfrac{s}{Sinh}\right)^{m-1}}{Sinh} - 1\right)\left(1 - \dfrac{s}{Sinh}\right)^{n}}{\left(s + Km\left(1 - \dfrac{s}{Sinh}\right)^{m}\right)^{2}} + \frac{mmax5nsx\left(1 - \dfrac{s}{Sinh}\right)^{n-1}}{SinhY\left(s + Km\left(1 - \dfrac{s}{Sinh}\right)^{m}\right)}\right] \right\}^{2}$$

$$\overline{\qquad Y\left(s + Km\left(1 - \dfrac{s}{Sinh}\right)^{m}\right) \qquad}$$

$$d = - \frac{mmax5s\left(1 - \dfrac{s}{Sinh}\right)^{n}}{Y\left(s + Km\left(1 - \dfrac{s}{Sinh}\right)^{m}\right)}$$

$$e = \frac{mmax5s\left(1-\dfrac{s}{Sinh}\right)^{n}\left[D+\dfrac{mmax5s\left(1-\dfrac{s}{Sinh}\right)^{n}}{Y\left(s+Km\left(1-\dfrac{s}{Sinh}\right)^{m}\right)}+\dfrac{mmax5sx\left(\dfrac{Kmn\left(1-\dfrac{s}{Sinh}\right)^{m-1}}{Sinh}-1\right)\left(1-\dfrac{s}{Sinh}\right)^{n}}{Y\left(s+Km\left(1-\dfrac{s}{Sinh}\right)^{m}\right)^{2}}-\dfrac{mmax5nsx\left(1-\dfrac{s}{Sinh}\right)^{m-1}}{SinhY\left(s+Km\left(1-\dfrac{s}{Sinh}\right)^{m}\right)}\right]}{Y\left(s+Km\left(1-\dfrac{s}{Sinh}\right)^{m}\right)}$$

Luong Kinetics

$$A=\begin{bmatrix} a & c \\ b & d \end{bmatrix}$$

$$a=-D-\frac{mmax6x\left(1-\dfrac{s}{Sh}\right)^{w}}{Y(Ks5+s)}+\frac{mmax6sx\left(1-\dfrac{s}{Sh}\right)^{w}}{Y(Ks5+s)^{2}}+\frac{mmax6wx\left(1-\dfrac{s}{Sh}\right)^{w-1}}{ShY(Ks5+s)}$$

$$b=\left[D+\frac{mmax6x\left(1-\dfrac{s}{Sh}\right)^{w}}{Y(Ks5+s)}-\frac{mmax6sx\left(1-\dfrac{s}{Sh}\right)^{w}}{Y(Ks5+s)^{2}}-\frac{mmax6x\left(1-\dfrac{s}{Sh}\right)^{w-1}}{ShY(Ks5+s)}\right]$$

$$+\frac{mmax6x\left(1-\dfrac{s}{Sh}\right)^{w}\left[\dfrac{mmax6x\left(1-\dfrac{s}{Sh}\right)^{w}}{(Ks5+s)^{2}}-\dfrac{mmax6sx\left(1-\dfrac{s}{Sh}\right)^{w}}{Y(Ks5+s)}-\dfrac{mmax6x\left(1-\dfrac{s}{Sh}\right)^{w-1}}{ShY(Ks5+s)}\right]}{Y(Ks5+s)}$$

$$c=-\frac{mmax6s\left(1-\dfrac{s}{Sh}\right)^{w}}{Y(Ks5+s)}$$

$$d=\frac{mmax6x\left(1-\dfrac{s}{Sh}\right)^{w}\left[D+\dfrac{mmax6x\left(1-\dfrac{s}{Sh}\right)^{w}}{Y(Ks5+s)}-\dfrac{mmax6sx\left(1-\dfrac{s}{Sh}\right)^{w}}{Y(Ks5+s)^{2}}-\dfrac{mmax6swx\left(1-\dfrac{s}{Sh}\right)^{w-1}}{ShY(Ks5+s)}\right]}{Y(Ks5+s)}$$

Study Case 6.6 Observability Matrix for Chemostat Model 3D (Product Inhibition) in Continuous Operation

μ = unstructured kinetic model considering substrate and product inhibition

$\mu(s)'$ = first derivate of μ respect to substrate

$\mu(p)'$ = first derivate of μ respect to substrate

$$A = \begin{bmatrix} -D - \dfrac{\mu(s)'x}{Y} & -\dfrac{\mu}{Y} & -\dfrac{\mu(p)'x}{Y} \\[2mm] \mu(s)'x & \mu - D & \mu(p)'x \\[2mm] \dfrac{\mu(s)'x}{Yp} & \dfrac{\mu}{Yp} & \dfrac{\mu(p)'x}{Y} - D \end{bmatrix} \quad ; \quad C = \begin{bmatrix} 1 \\ 0 \\ 0 \end{bmatrix}_T \ ;$$

rank(A)=3 ∴The system is invertible

$$Ob = \begin{bmatrix} 1, & 0 & 0 \\[3mm] -D - \dfrac{\mu(s)'x}{Y}, & -\dfrac{\mu(s)}{Y} & -\dfrac{\mu(p)'x}{Y} \\[3mm] \left(D + \dfrac{\mu(s)'x}{Y}\right)^2 - \dfrac{\mu(s)\mu(s)'x}{Y} - \dfrac{\mu(p)'\mu(s)'x^2}{YYp}, & \dfrac{\mu(s)\left(D + \dfrac{\mu(s)'x}{Y}\right)}{Y} + \dfrac{\mu(s)(D - \mu(s))}{Y} - \dfrac{\mu(p)'\mu(s)x}{YYp}, & \dfrac{\mu(p)'x\left(D + \dfrac{\mu(s)'x}{Y}\right)}{Y} + \dfrac{\mu(p)'\mu(s)x}{Y} - \dfrac{\mu(p)'x\left(D + \dfrac{\mu(p)'x}{Yp}\right)}{Y} \end{bmatrix}$$

rank(Ob)=2 ∴The system is not observable

$Det(Ob) = 0$

If $\mu(s) = D$ in the stationary state so,

$$A = \begin{bmatrix} -D - \dfrac{\mu(s)'x}{Y} & -\dfrac{\mu(s)}{Y} & -\dfrac{\mu(p)'x}{Y} \\[2mm] \mu(s)'x & 0 & \mu(p)'x \\[2mm] \dfrac{\mu(s)'x}{Yp} & \dfrac{\mu(s)}{Yp} & \dfrac{\mu(p)'x}{Y} - D \end{bmatrix} \quad ; \quad C = \begin{bmatrix} 1 \\ 0 \\ 0 \end{bmatrix}_T \ ;$$

rank(A)=3 ∴The system is invertible

$$Ob = \begin{bmatrix} 1, & 0 & 0 \\[2mm] -D - \dfrac{\mu(s)'x}{Y}, & -\dfrac{\mu(s)}{Y} & -\dfrac{\mu(p)'x}{Y} \\[3mm] \left(D + \dfrac{\mu(s)'x}{Y}\right)^2 & \mu(s)\left(D + \dfrac{\mu(s)'x}{Y}\right) & \mu(p)'x\left(D + \dfrac{\mu(s)'x}{Y}\right) \\ -\dfrac{\mu(s)\mu(s)'x}{Y} - \dfrac{\mu(p)'\mu(s)'x^2}{YYp}, & -\dfrac{\mu(s)\left(D + \dfrac{\mu(s)'x}{Y}\right)}{Y} - \dfrac{\mu(p)'\mu(s)x}{YYp}, & -\dfrac{\mu(p)'\mu(s)x}{Y} + \dfrac{\mu(p)'x\left(D + \dfrac{\mu(p)'x}{Yp}\right)}{Y} \end{bmatrix}$$

rank(Ob)=3 ∴The system is observable

$$Det(Ob) = \dfrac{\mu(s)^2\,\mu(p)'x - D\mu(s)\mu(p)'x}{Y^2}, \text{ but } \mu(s) = D$$

∴ $Det(Ob) = 0 \rightarrow$ *the system is in fact not observable*

The observability matrix was proven with another output as $C = \begin{bmatrix}0 & 1 & 0\end{bmatrix}$; $C = \begin{bmatrix}0 & 0 & 1\end{bmatrix}$; $C = \begin{bmatrix}1 & 0 & 1\end{bmatrix}$; $C = \begin{bmatrix}1 & 1 & 0\end{bmatrix}$; *and* $C = \begin{bmatrix}0 & 1 & 1\end{bmatrix}$ and we observed that the system is still no observable.

Study Case 6.7 Observability Matrix for Chemostat Model 3D (Product Inhibition) in Batch Operation

$$A = \begin{bmatrix} -\dfrac{\mu(s)'x}{Y} & -\dfrac{\mu}{Y} & -\dfrac{\mu(p)'x}{Y} \\[3mm] \mu(s)'x & \mu & \mu(p)'x \\[3mm] \dfrac{\mu(s)'x}{Yp} & \dfrac{\mu}{Yp} & \dfrac{\mu(p)'x}{Y} \end{bmatrix} \quad ; \quad C = \begin{bmatrix} 1 \\ 0 \\ 0 \end{bmatrix}_T \; ;$$

rank(A)=1 ∴ The system is not invertible

$$Ob = \begin{bmatrix}
1 & -\dfrac{\mu(s)'x}{Y} & 0 \\[3mm]
-\left(\dfrac{\mu(s)'x}{Y}\right)^2 & -\dfrac{\mu(s)}{Y} & -\dfrac{\mu(p)'x}{Y} \\[3mm]
-\dfrac{\mu(s)\mu(s)'x}{Y} - \dfrac{\mu(p)'\mu(s)'x^2}{YYp} & \dfrac{\mu(s)^2}{Y} + \dfrac{\mu(p)'\mu(s)'x}{YYp} & \dfrac{\mu(p)'\mu(s)'x^2}{Y^2} + \dfrac{\mu(p)'^2 x^2}{YYp} - \dfrac{\mu(s)\mu(p)'x}{Y}
\end{bmatrix}$$

$$\dfrac{\mu(s)\mu(s)'x}{Y} + \dfrac{\mu(s)^2}{Y} - \dfrac{\mu(p)'\mu(s)'x}{YYp},$$

rank(Ob) = 2 ∴ The system is not observable $Det(Ob) = 0$. The observability matrix was proven with another output as $C = [0\ 1\ 0]$; $C = [0\ 0\ 1]$; $C = [1\ 0\ 1]$; $C = [1\ 1\ 0]$; and $C = [0\ 1\ 1]$ and we observed that the system is still not observable.

Remarks   The following (Table 6.2) contains a summary of the observability tests considering different inputs and operation cases, with the purpose of capturing process states for monitoring purposes, and to prevent bioprocessing disruptions. Unfortunately, not all of the important states are measurable due to the lack of appropriate devices and feasibility of design and implementation sensors in the bioprocesses.

Table 6.2   Overall results of the observability analysis applied to 2D and 3D bioreactor.

System	Case-operation	Output	R	Observability Yes	No
2D	1. Batch	$C = \begin{bmatrix} 1 & 0 \end{bmatrix}$	2	↑	
	2. Batch cell death	$C = \begin{bmatrix} 1 & 0 \end{bmatrix}$	2	↑	
	3. Batch yield not constant	$C = \begin{bmatrix} 1 & 0 \end{bmatrix}$	2	↑	
3D	4. Batch product inhibition	$C = \begin{bmatrix} 1 & 0 & 0 \end{bmatrix}$	1		↓
		$C = \begin{bmatrix} 0 & 1 & 0 \end{bmatrix}$	2		↓
		$C = \begin{bmatrix} 1 & 0 & 1 \end{bmatrix}$	2		↓
		$C = \begin{bmatrix} 1 & 1 & 0 \end{bmatrix}$	2		↓
		$C = \begin{bmatrix} 0 & 1 & 1 \end{bmatrix}$	2		↓
2D	5. Continuous	$C = \begin{bmatrix} 1 & 0 \end{bmatrix}$	2	↑	
	6. Continuous cell death	$C = \begin{bmatrix} 1 & 0 \end{bmatrix}$	2	↑	
3D	7. Continuous product inhibition	$C = \begin{bmatrix} 1 & 0 & 0 \end{bmatrix}$	2		↓
		$C = \begin{bmatrix} 0 & 1 & 0 \end{bmatrix}$	2		↓
		$C = \begin{bmatrix} 1 & 0 & 1 \end{bmatrix}$	2		↓
		$C = \begin{bmatrix} 1 & 1 & 0 \end{bmatrix}$	2		↓
		$C = \begin{bmatrix} 0 & 1 & 1 \end{bmatrix}$	2		↓

6.4 Implementation of a Linear Observer to Check the Results of the Observability Analysis

For validated purposes, a Luenberger observer was implemented, with the following initial conditions for the virtual plant (model) and the observers: $X_0 = [8 \ 0.9 \ 1.5]^T$ and $\hat{x}_0 = [9 \ 1.3 \ 1.0]^T$.

In this chapter the concept of observability has been shown, which is well understood for continuous and batch processes, as is illustrated in Table 6.2. Therefore, the system is observable if and only if the observability matrix of the system is full rank [11–15], otherwise the system is unobservable; this means the current values of some of its states cannot be determined through output sensors [16–21]. Hence the system (6.1)–(6.12) is not fully observable for 3D systems in batch and continuous but for 2D is observable in batch and continuous operations.

The observer provides a good state estimation, the observer gain is $K = 10 \, \text{h}^{-1}$. The trajectories of the proposed methodology converge quickly to the real trajectories. Figure 6.1 shows the comparison of real data with the estimated values for substrate, biomass, and

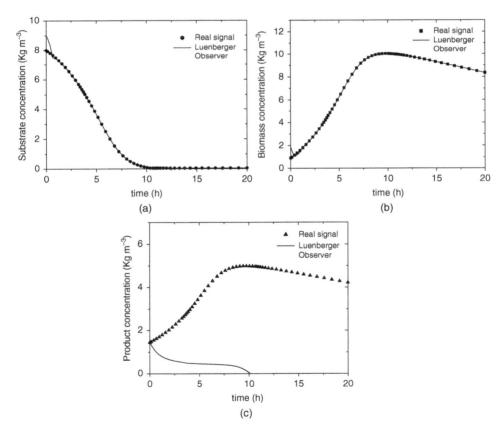

Figure 6.1 Results of (symbols) actual values, (lines) Luenberger observer. Case 3D in continuous product inhibition.

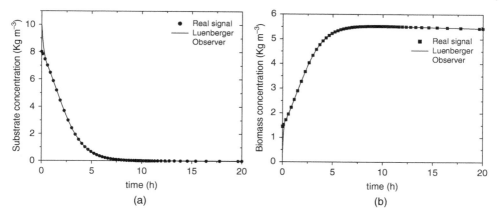

Figure 6.2 Results of (symbols) actual values, (lines) Luenberger observer. Case 2D in continuous product inhibition.

product concentrations for 3D in conditions of continuous product inhibition D = 0.01 h^{-1}. These results validate the observability of the system found. Figure 6.2 shows the comparison of real data with the estimated values for substrate, and biomass concentrations for 2D in conditions of continuous and Monod kinetics and D = 0.01 h^{-1}. These results validate the observability of the system found.

From a mass balance for the substrate and biomass in the bioreactor (6.1, 6.2), the following state observer is obtained:

$$\frac{d\widehat{S}}{dt} = DS_i - D\widehat{S} - \mu_{max}\frac{\widehat{S}}{k_s\widehat{S}}\frac{\widehat{X}}{Y} + \overbrace{k_l}^{\text{observer gain}}(S - \widehat{S}) \tag{6.34}$$

$$\frac{dX}{dt} = -DX - \mu_{max}\frac{\widehat{S}}{k_s\widehat{S}}\frac{\widehat{X}}{Y} + k_l(S - \widehat{S}) \tag{6.35}$$

6.5 Conclusion

The present work is aimed at providing a solution of the measurement problems in bioreactors with respect to the biological variables, on the basis of only one relatively easilt measurable variable, thus creating a firm basis for advanced control strategies. Several output measurement combinations were proposed in order to show the dimensions of the corresponding observable subspaces employing the observability matrix criterion. The chemostat in 2D (batch or continuous) will always be observable with the only condition that the growth rate is non-zero. The chemostat in 3D is non-observable no matter whether the reactor is operated in batch or continuous.

List of Figures

References

1 Nielsen, J.H., Villadsen, J., and Liden, G. (2003). *Bioreaction Engineering Principles*. New York: Springer.

2 Fu, G. and Ma, W. (2006). Hopf bifurcations of a variable yield chemostat model with inhibitory exponential substrate uptake. *Chaos, Solitons & Fractals* 30: 845.

3 Dong, Q.L. and Ma, W.B. (2013). Qualitative analysis of the chemostat model with variable yield and a time delay. *Journal of Mathematical Chemistry* 51: 1274.

4 López-Pérez, P.A., Maya-Yescas, R., Peña-Caballero, V. et al. (2013). Software sensors design for a model of a simultaneous saccharification and fermentation of starch to ethanol. *Fuel* 110: 219–226.

5 Dang, T., Le Guernic, C., and Maler, O. (2011). Computing reachable states for nonlinear biological models. *Theoretical Computer Science* 412: 2095–2107.

6 Vidal, R., Chiuso, A., and Soatto, S. (2002). Observability and identifiability of jump linear systems. In: *Proceedings of the 41st IEEE Conference on Decision and Control*, 3614–3619. IEEE.

7 Morari, M. and Stephanopoulos, G. (1980). Studies in the synthesis of control structures for chemical processes: part II: structural aspects and the synthesis of alternative feasible control schemes. *AICHE Journal* 26: 232–246.

8 López-Pérez, P.A., Cuevas, F.A., Gomez, R.V. et al. (2015). Improving bioethanol production via nonlinear controller with noisy measurements. *Chemical Engineering Communications* 202 (11): 1438–1445.

9 Agarwal, R., Mahanty, B., and Dasu, V.V. (2009). Modeling growth of *Cellulomonas cellulans* nrrl b 4567 under substrate inhibition during cellulase production. *Chemical and Biochemical Engineering Quarterly* 23: 213.

10 López-Pérez, P.A., Puebla, H., Velázquez Sánchez, H.I. et al. (2017). Comparison tools for parametric identification of kinetic model for ethanol production using evolutionary optimization approach. *International Journal of Chemical Reactor Engineering* 14 (6).

11 Sterner, R.W., Small, G.E., and Hood, J.M. (2012). The conservation of mass. *Nature Education Knowledge* 3: 20.

12 Gauthier, J., Hammouri, H., and Othman, S. (1992). A simple observer for nonlinear systems applications to bioreactors. *IEEE Transactions on Automatic Control* 37: 875–880.

13 Dochain, D. and Chen, L. (1992). Local observability and controllability of stirred tank reactors. *Journal of Process Control* 2: 139–144.

14 Moreno, J.A. Observer design for bioprocesses using a dissipative approach. 17th IFAC World Congress, Seoul, Korea, July 6-11, 2008

15 Martinez-Guerra, R., Garrido, R., and Osorio-Miron, A. (2001, 5). Parametric and state estimation by means of high-gain nonlinear observers: application to a bioreactor. In: *Proceedings of the 2001 American Control Conference. (Cat. No.01CH37148), Arlington, VA, USA*, 3807–3808.

16 De Villeros, P., Botero, H., and Alvarez, H. (2016). State observer design for biomass and ethanol estimation in bioreactors using cybernetic models. *DYNA* 83 (198): 119–127.

17 Lillacci, G. and Valigi, P. (2007). State observers for the estimation of mRNA and protein dynamics. In: *2007 IEEE/NIH Life Science Systems and Applications Workshop*, 108–111.

18 Waldherr, S. (2018). Estimation methods for heterogeneous cell population models in systems biology. *Journal of the Royal Society Interface* 15.

19 Dürr, R. and Steffen, W. (2018). A novel framework for parameter and state estimation of multicellular systems using Gaussian mixture approximations. *Processes* 6 (10): 187.

20 Fox, Z.R. and Munsky, B. (2019). The finite state projection based fisher information matrix approach to estimate information and optimize single-cell experiments. *PLoS Computational Biology* 15 (1): e1006365.

21 Aguilar-Lopez, R. and Mata-Machuca, J. (2018). On the observability and state estimation in a class of gene-expression system. *Dynamic Systems and Applications* 27: 531–544.

7

Production System Myco-Diesel for Implementation of "Quality" of the Observability

The production of myco-diesel hydrocarbons by the endophytic fungi is a new alternative and environmentally sustainable source. When compared to traditional fossil fuels it has reduced gas emissions and is made from renewable bioprocess sources such as vegetable oils, animal fats, and microorganism cultures. The biodiesel is biodegradable and non-toxic, in essence by a set of monoalkyl esters of long-chain fatty acids, which at present is derived chiefly from the acylglycerols of vegetable oils by transesterification or esterification with short chain alcohols. Currently, the principal plant oils employed are oil crops such as palm, oilseed rape, and soybean. This chapter shows that the observability property (conditions for state estimation)in the mycodiesel bioreactor model, as a study case, is affected by the equilibrium point chosen and the number of measurable outputs, this was explained by the observability matrix construction and its singular values and condition number. For the mycodiesel bioreactor in continuous operation, the nitrogen dynamic is not observable but at least detectable both biomass and biomass-substrate as output measurements, however in batch operation the system is neither observable or detectable considering the same couple of outputs.

7.1 Introduction

The production of fuel-diesel from biological systems can avoid the production of environmental greenhouse gasses caused by straw waste incineration and can play an important role in realizing sustainable economic development and ecological environment protection. The process of straw fermentation to produce fuel ethanol is a highly nonlinear, large time-delay and multivariable dynamic coupling process. Moreover, in the straw fermentation process, some crucial parameters (bacteria concentration, fermentable substrate, and fuel concentration) cannot be measured online in real time and there is no accurate mechanism model to be used.

Biodiesel is defined as a mixture of fatty acid alkyl esters, mainly C16 and C18, which can be produced from renewable sources. It is biodegradable, non-toxic, has a low potential explosive risk, and has similar properties to diesel. The raw materials used for the production of biodiesel are vegetable and animal oils or fats, and waste cooking oils, but there is another alternative, i.e. microorganisms. Microorganisms that are capable of accumulating single cell oils (SCO) in more than 20% of their dry cell weight are called

Control in Bioprocessing: Modeling, Estimation and the Use of Soft Sensors, First Edition.
Pablo Antonio López Pérez, Ricardo Aguilar López, and Ricardo Femat.
© 2020 John Wiley & Sons Ltd. Published 2020 by John Wiley & Sons Ltd.

oleaginous organisms [1]. Under conditions of nitrogen limitation these microorganisms can increase their lipids in the form of triacylglycerols and through a transesterification reaction biodiesel is produced.

The biodiesel obtained from oleaginous fungus is named mycodiesel, it shows advantages over the lipids obtained from crop plants because it doesn't compete with food supply usage; has better cold flow properties than bacteria; doesn't require large acreages or artificial lightening as microalgae, nevertheless, the cost of biodiesel is still very high, so it is necessary to find strategies to improve the process economics. To achieve the latter, different methodologies have been proposed in order to increase the lipid content, the productivity and lower the process cost. For example, experimental studies have been carried out that changed the temperature, pH and carbon/nitrogen radios [2–4]; fermentations with renewable carbon sources [5–9]; using different substrate feeding strategies [10, 11]; mathematical modeling, sensitivity, and dynamical analysis of lipid biosynthesis for the biofuel production [12–17]; optimization of oil productivity [18–20]; state estimation and control of biodiesel process [21–23].

Crucial parameters have now been obtained through offline analysis and laboratory tests; however, there is a big time-delay in data measurement, and thus it is difficult to meet real-time control requirements for the straw fermentation process. Therefore, research is required to obtain crucial status information of bioprocesses which is important to optimize the straw fermentation process and thus improving fuel yield and quality.

This work discusses the variations in the local observability/detectability property for a biodiesel bioreactor model in batch and continuous operation, by changes in the operation conditions (the dilution rate) and the measurable outputs C1 (biomass) and C2 (biomass-substrate). This analysis is supported by obtaining the singular value decomposition (SVD) and the condition number (CN) for the observability matrix, in order to explain the weak, or loss of, observability. The observability/detectability results are validated by simulations with the implementation of an extended Luenberger observer for the estimation of lipids in the biodiesel bioreactor model for both C1 and C2; finally, the state estimation performance is measured by the ITSE.

7.2 Methodology

7.2.1 Local Observability Quality

In relation to the bioprocess state estimation, it is well known that for a successful observer implementation it has to be verified that the process is observable.

The observability is commonly calculated locally, i.e. on a specific equilibrium point of interest, but that this system property varies with the process input/output data considering the number, and location of sensors and the mathematical model that describes the process, has been quite well studied. There are only a few studies to our knowledge related to this, such as [24]. In this work the optimal sensor location for nonlinear dynamic systems by using empirical observability gramians for observability analysis without resorting to linearization of the model was determined. In [25], the practical (numerical) observability matrix by the SVD and the conditioning number required to determinate the best observability possible in practice for a heterogeneous bioreactor in continuous operation was analyzed where the effect of 1, 2, and 3 sensors at different locations in the bioreactor was considered. In [26] supported by the observability matrix, the observability gramians and the Popov–Belevitch–Hautus rank test, the optimal sensor positions for a tubular reactor with a preset number of sensors were determined.

7.2.2 Bioreactor Model

Oleaginous yeasts can be produced in batch(intermittent), fed-batch and continuous mode. The drawback of the batch mode is the unfeasibility of regulating the C/N ratio in the bioreactor, whereas a fed-batch strategy is effective for regulating the C/N ratio, finally, operation in a continuous mode does not allow the regulation of the nutrient flow rate in the bioprocess (C/N ratio). However, if n inputs flow rates are considered, C/N regulation is possible. The above justifies the analysis of the different inputs for the detectability, mainly the operation in continuous mode [27].

The bioreactor model corresponds to the lipid production by the oleaginous fungus *Mortierella isabellina*. It considers inhibition at high sugar concentrations. The model parameters are shown in Table 7.1 and Figure 7.1. The mass equation balances in the bioreactor are as follows [28]:

For fat-free biomass (X):

$$\frac{dX}{dt} = -DX + \mu_{SN}X \tag{7.1}$$

For storage lipids (L):

$$\frac{dL}{dt} = -DL + q_L X \tag{7.2}$$

Table 7.1 Kinetic parameters for *Mortierella isabellina* growth.

Parameters	Value	Definition	Units
μ_{SNmax}	0.566 ± 0.1	Maximal growth rate on carbon and nitrogen	h^{-1}
Y_{XS}	0.345 ± 0.10	The yield of fat-free biomass with respect to carbon substrate	$g\,g^{-1}$
Y_{LS}	18.209 ± 0.150	Yield lipids on biomass	$g\,g^{-1}$
K_S	0.615 ± 0.1	Carbon saturation constant for growth on carbon and nitrogen	$g\,l^{-1}$
Y_{XN}	18.209 ± 3	The yield of fat-free biomass with respect to nitrogen	$g\,l^{-1}$
K_{LS}	8.1355 ± 1	The yield of lipid biomass with respect to carbon substrate	$g\,l^{-1}$
K_N	1.5885 ± 0.1	Nitrogen saturation constant for growth on carbon and nitrogen	$g\,l^{-1}$
k_2	13.631 ± 4	Constant of nitrogen regulation for lipid production	$g\,l^{-1}$
K_{i1}	8.5139 ± 1	Inhibition constant for growth on carbon and nitrogen	$g\,l^{-1}$
K_{i2}	0.395 ± 0.1	Inhibition constant for growth on carbon and nitrogen	$g\,l^{-1}$
q_{Lmax}	0.785 ± 0.050	Productivity	$g\,l^{-1}\,h$
V	1.0	Volume of the reactor	l
D	$0{-}1$	Dilution rate	h^{-1}

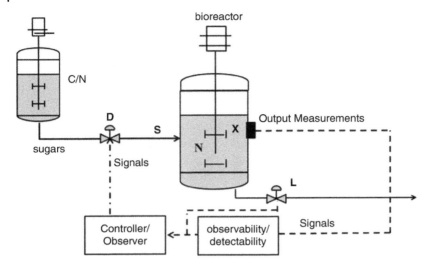

Figure 7.1 Bioreactor scheme for continuous culture system.

For the hydrolyzed sugars (S):

$$\frac{dS}{dt} = D(S_f - S) - \left(\frac{\mu_{SN}}{Y_{XS}} + \frac{q_L}{Y_{LS}}\right)X \qquad (7.3)$$

For nitrogen consumption (N):

$$\frac{dN}{dt} = D(N_f - N) - \frac{\mu_{SN}}{Y_{XN}}X \qquad (7.4)$$

Kinetic growth model for double substrate consumption

$$\mu_{SN}(S, N) = \mu_{SNmax}\frac{S}{K_S + S + \frac{S^2}{K_{i1}}}\frac{N}{K_N + N} \qquad (7.5)$$

Consumption rate, $q_L(S, N)\, q_L(S, N)$:

$$q_L(S, N) = q_{Lmax}\frac{S}{K_{LS} + S + \frac{S^2}{K_{i2}}}\frac{k_2}{k_2 + N} \qquad (7.6)$$

For batch operation, $D = 0$.

The definition, units, and values of the parameters of the model are listed in Table 7.1. They were obtained from batch cultures of *M. isabellina* on sweet sorghum as a carbon substrate [28]. For now, this set of parameters are the initial parameters for the simulations.

7.3 Main Results

As mentioned before, the model parameters correspond to a batch culture with *M. isabellina*, [28]. Also, the coefficient describing the regulation of nitrogen to the lipid production (k_2) has a high value, that is $k_2 \gg N$ which implies that under the experimental conditions of the culture, the production of lipids is weakly regulated by the nitrogen.

For instance, *Yarrowia lipolytica,* when grown on industrial fats, was able to utilize its accumulated lipids for the synthesis of fat-free biomass [29].

The numerical simulations for the bifurcation analysis of the bioreactor model were done in Matcont v.5.0, a free MATLAB package for numerical bifurcation analysis of ODE mathematical models. Some particular equilibrium points were chosen from the above bifurcation analysis to illustrate their trajectories, attraction domains, and stabilities, through the construction of phase portraits, using pplane8, a MATLAB package for numerical analysis of ODEs [30].

The main objective of bifurcation theory is to characterize changes in the qualitative dynamic behavior of a nonlinear system as the key parameter values are changed, for example, dilution rate, parameter kinetics, and time delay in the bioprocess. This means that the system achieves a critical value, where an orbit change occurs and, as a consequence, the possibility of different stability properties of equilibrium, the system is structurally stable. Properly, bifurcation can be introduced as the appearance of a topologically nonlinear phase portrait under variation of a parameter [31, 32].

A numerical bifurcation of the mycodiesel bioreactor model was done previously to the observability analysis in order to show the model qualitative behavior, stability, and the operating conditions required to yield high lipid productivity. The dilution rate was taken as the bifurcation parameter for a high C/N ratio value of 90. It was reported that high C/N ratios induce lipogenesis [4].

Figure 7.2 (left side) shows the bifurcation diagrams for each unstructured model. Some critical points are marked on the diagrams; these are the branch point 1 (BP1), branch point 2 (BP2) and the limit point (LP), which represent the biomass concentration in batch culture, washout condition (trivial solution) and maximum operating dilution rate, respectively. The equilibrium points from BP1 to LP are stable nodes, while

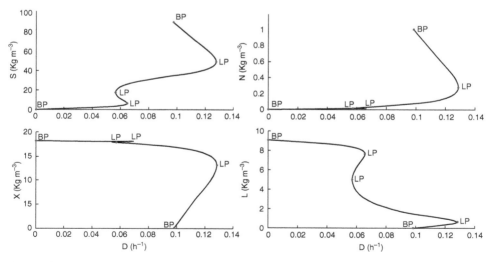

Figure 7.2 Bifurcation diagram of the four state variables: hydrolyzed sugar concentration (S), nitrogen concentration (N), biomass concentration (X) and lipid concentration (L). Considering the dilution rate (D) as the bifurcation parameter.

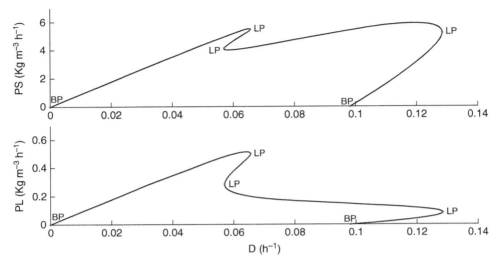

Figure 7.3 Bifurcation diagram of the sugar consumption (PS) and lipid production (PL) rates considering the dilution rate (D) as the bifurcation parameter.

those from LP to BP2 are saddle points, this last interval corresponds to a multiplicity steady-state region.

The numerical simulations for the bifurcation analysis of the proposed variable dilution rate model is a realistic approach to describe the behavior of the cells and can be extrapolated to several operating conditions. The analysis shows rich dynamical behavior from a multiplicity of equilibria to different bifurcation types for the bioreactor model with the proposed approach.

Figure 7.2 shows the equilibrium branches for biomass, substrate, lipid, and nitrogen concentrations. It can be noted that the mycodiesel bioreactor model has a multiplicity of steady states for two intervals, close to the values 0.06 and 0.12 h^{-1}. For the first value a triple steady state multiplicity appears in which two equilibrium points are stable nodes and another is an unstable node; in the second value, a stable with an unstable node coexist.

Figure 7.3 shows the lipid production and substrate consumption velocities. There are two equilibrium points where the maximum productivity is found, at 0.06 h^{-1} and close to the washout dilution rate (0.12 h^{-1}), the first point was taken as the better condition for lipid production because of its high lipid productivity and substrate consumption rate, so for this equilibrium a state observer for lipid estimation is required.

The observability and detectability analysis were done by the observability matrix and the Popov–Belevitch–Hautus (PBH) frameworks for all the equilibrium branches, considering C_1 and C_2 as the most suitable sets of measurable outputs. The observability matrix for the batch bioreactor, taking into account both output cases, showed an analytical rank of $O_{ba} = 3O_{ba} = 3$, however, the numerical rank was $O_{be} = 2$. The PBH theorem shows a detectability rank of $O_{bd} = 3O_{bd} = 3$, so the batch bioreactor is not detectable. For the continuous bioreactor, the observability analytical rank was $O_{ca} = 3O_{ca} = 3$, and a numerical rank of $O_{ce} = 3O_{ce} = 3$ (data not shown). The detectability rank was $O_{cd} = 4O_{cd} = 4$, hence the

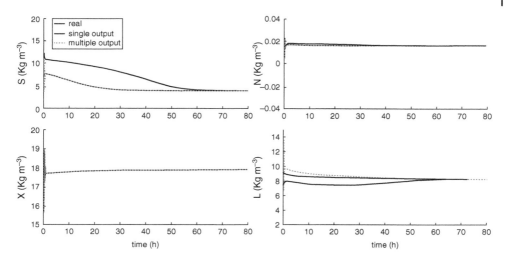

Figure 7.4 Variations in the singular values of the observability matrix with single (biomass) and multiple (biomass-substrate) measurable outputs.

continuous reactor is detectable. The above results are the same throughout the operating bioreactor dilution rate interval (see Figure 7.4).

Theorem 7.1 The pair (A, C) is observable if and only if there exists no $x \neq 0$ such that

$$Ax = \lambda x, \quad \forall \quad Cx = 0 \tag{7.7}$$

Detectability

Problem: Given $y(t)$ over $t \in [0, t]$ with $t > 0$ can one compute $x(t)$?
   Theorem: The following are equivalent

a) The pair (A, C) is detectable;
b) There exists no $x \neq 0$ and λ such that $Ax = \lambda x$, $Cx = 0$ with $\lambda + \lambda^* \geq 0$.

The relationship between controlled and estimated, measured (input) and manipulated variables (output) and how to link these variables to form control loops systems are problems that, in bioprocess practice, are solved heuristically, existing methods for control structure design are based on relative gains and singular value analysis [33, 34]. The control structure design criteria are based in the maximum singular value, the condition number and the singular vectors:

- Maximum singular values: In general, it is desirable that the maximum singular value be small. In Havre et al. (1996) and Skogestad and Postlethwaite (1996), this index is used as a criterion for selecting secondary measurements, an SVD analysis of the transfer functions that relate the output error with the disturbance and the uncertainty is carried out.
- Minimum singular value: Morari (1983), Skogestad and Postlethwaite (1996), argue that this value should be big in order for a plant to have a good tracking and regulation performance, independent control of the variables can be guaranteed.

- The SVD and the condition number are shown in Figures 7.6 and 7.7, it can be noted that this values changes with the equilibrium point chosen, for the mycodiesel model the highest values are near the washout dilution rate, which means that the observability/detectability is diminished.

A Luenberger observer was implemented at $0.06\,\mathrm{h^{-1}}$, taking into account the two output measurable cases in order to validate the above results. Simulations show the estimated trajectory and the performance index as the ITSE [35] (Figures 7.5–7.6). The gain values ($k_1 = 1.5\,\mathrm{h^{-1}}$) agree with the operating conditions. The observer performance was improved employing multiple outputs instead of a single output. Finally, the results obtained here can

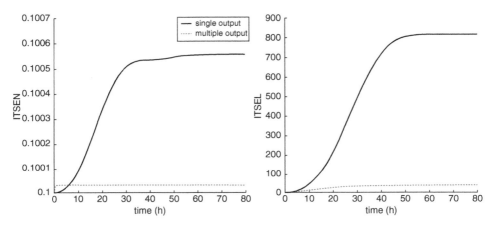

Figure 7.5 Condition number for the observability matrix (logarithmic scale) at different dilution rates for simple and multiple outputs.

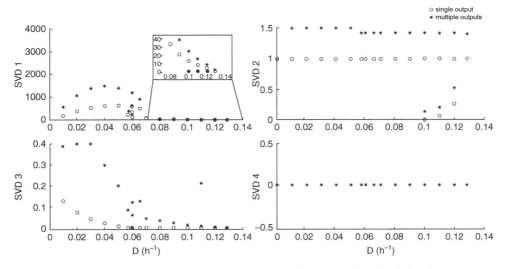

Figure 7.6 Simulation of a continuous bioreactor that produces mycodiesel and the observer performance with a single and multiple outputs.

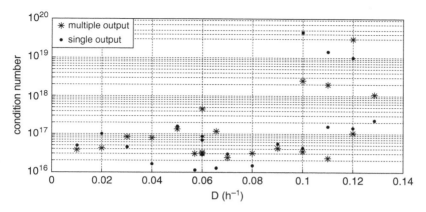

Figure 7.7 Performance index (ITSE) of the observer for the nitrogen concentration (ITSE N) and lipid concentration (ITSE L) state variables.

help to explain the low performance of some control based on estimators because of low observability property in the equilibrium point of interest.

The observer has a different set of initial conditions from the real plant simulation. The state variables are hydrolyzed sugar concentration (S), nitrogen concentration (N), biomass concentration (X), and lipid concentration (L).

The closed-loop state and input observer with small performance (condition number) indices are considered as a well-stable input observer (Figure 7.7).

Actually, average productivity of the industrial strains is far lower than real theoretical estimations, suggesting that identification of factors limiting bioprocess (biofuels) yield, instrumentation and control are crucial in the industrial scale [36–40].

7.4 Conclusions

This study shows that the observability property in the mycodiesel bioreactor model, as a study case, is affected by the equilibrium point chosen and the number of measurable outputs, this was explained by the observability matrix construction and its singular values and condition number. For the mycodiesel bioreactor in continuous operation, the nitrogen dynamic is not observable but at least detectable both biomass and biomass-substrate as output measurements, however in batch operation the system is neither observable or detectable considering the same couple of outputs.

List of Figures

List of Table

References

1 Li, N., Deng, Z.N., Qin, Y.L. et al. (2008). Production of polyunsaturated fatty acids by *Mucor recurvus* sp. with sugarcane molasses as carbon source. *Food Technology and Biotechnology* 46 (1): 73–79.

2 Angerbauer, C., Siebenhofer, M., Mittelbach, M. et al. (2008). Conversion of sewage sludge into lipids by *Lipomyces starkeyi* for biodiesel production. *Technology* 99: 3051–3056.

3 Economou, C.N., Aggelis, G., Pavlou, S. et al. (2011). Single cell oil production from rice hulls hydrolysate. *Bioresource Technology* 102: 9737–9742.

4 Subhash, G.V. and Mohan, S.V. (2014). Lipid accumulation for biodiesel production by oleaginous fungus *Aspergillus awamori*: influence of critical factors. *Fuel* 116: 509–515.

5 Fakas, S., Papanikolaou, S., Batsos, A. et al. (2009). Evaluating renewable carbon sources as substrates for single cell oil production by *Cunninghamella echinulata* and *Mortierella isabellina*. *Biomass and Bioenergy* 33: 573–580.

6 Zhao, X., Kong, X.L., Hua, Y.Y. et al. (2008). Medium optimization for lipid production through co-fermentation of glucose and xylose by the oleaginous yeast *Lipomyces starkeyi*. *European Journal of Lipid Science and Technology* 110: 405–412.

7 Economou, C.N., Makri, A., Aggelis, G. et al. (2010). Semi-solid state fermentation of sweet sorghum for the biotechnological production of single cell oil. *Bioresource Technology* 101: 1385–1388.

8 Papanikolaou, S., Diamantopoulou, P., Chatzifragkou, A. et al. (2010). Suitability of low-cost sugars as substrates for lipid production by the fungus *Thamnidium elegans*. *Energy & Fuels* 24: 4078–4086.

9 Gong, Z.W., Wang, Q., Shen, H.W. et al. (2012). Co-fermentation of cellobiose and xylose by *Lipomyces starkeyi* for lipid production. *Bioresource Technology* 117: 20–24.

10 Zhao, X., Hu, C.M., Wu, S.G. et al. (2011). Lipid production by *Rhodosporidium toruloides* Y4 using different substrate feeding strategies. *Journal of Industrial Microbiology & Biotechnology* 38: 627–632.

11 Fei, Q., Chang, H.N., Shang, L., and Choi, J.D.R. (2011). Exploring low-cost carbon sources for microbial lipids production by fed-batch cultivation of *Cryptococcus albidus*. *Biotechnology and Bioprocess Engineering* 16: 482–487.

12 Meeuwse, P., Akbari, P., Tramper, J., and Rinzema, A. (2012). Modeling growth, lipid accumulation and lipid turnover in submerged batch cultures of *Umbelopsis isabellina*. *Bioprocess and Biosystems Engineering* 35: 591–603.

13 Meeuwse, P., Tramper, J., and Rinzema, A. (2011). Modeling lipid accumulation in oleaginous fungi in chemostat cultures: I. Development and validation of a chemostat model for *Umbelopsis isabellina*. *Bioprocess and Biosystems Engineering* 34: 939–949.

14 Meeuwse, P., Tramper, J., and Rinzema, A. (2011). Modeling lipid accumulation in oleaginous fungi in chemostat cultures. II: validation of the chemostat model using yeast culture data from literature. *Bioprocess and Biosystems Engineering* 34: 951–961.

15 Shen, H.W., Gong, Z.W., Yang, X.B. et al. (2013). Kinetics of continuous cultivation of the oleaginous yeast *Rhodosporidium toruloides*. *Journal of Biotechnology* 168: 85–89.

16 Tevatia, R., Demirel, Y., and Blum, P. (2012). Kinetic modeling of photoautotropic growth and neutral lipid accumulation in terms of ammonium concentration in *Chlamydomonas reinhardtii*. *Bioresource Technology* 119: 419–424.

17 Surisetty, K., Siegler, H.D., McCaffrey, W.C., and Ben-Zvi, A. (2010). Robust modeling of a microalgal heterotrophic fed-batch bioreactor. *Chemical Engineering Science* 65: 5402–5410.

18 Siegler, H.D., McCaffrey, W.C., Burrell, R.E., and Ben-Zvi, A. (2012). Optimization of microalgal productivity using an adaptive, nonlinear model based strategy. *Bioresource Technology* 104: 537–546.

19 Simionato, D., Basso, S., Giacometti, G.M., and Morosinotto, T. (2013). Optimization of light use efficiency for biofuel production in algae. *Biophysical Chemistry* 182: 71–78.

20 El-Sheekh, M., Abomohra, A., and Hanelt, D. (2013). Optimization of biomass and fatty acid productivity of *Scenedesmus obliquus* as a promising microalga for biodiesel production. *World Journal of Microbiology and Biotechnology* 29: 915–922.

21 Abdollahi, J. and Dubljevic, S. (2012). Lipid production optimization and optimal control of heterotrophic microalgae fed-batch bioreactor. *Chemical Engineering Science* 84: 619–627.

22 Nadadoor, V.R., Siegler, H.D., Shah, S.L. et al. (2012). Online sensor for monitoring a microalgal bioreactor system using support vector regression. *Chemometrics and Intelligent Laboratory Systems* 110: 38–48.

23 Havlik, I., Lindner, P., Scheper, T., and Reardon, K.F. (2013). On-line monitoring of large cultivations of microalgae and cyanobacteria. *Trends in Biotechnology* 31: 406–414.

24 Singh, H.J. (2005). Optimal sensor location for nonlinear dynamic systems via empirical Gramians. In: *Dynamics and Control of Process Systems 2004 (DYCOPS-7): A Proceedings Volume from the 7th IFAC Symposium*, Cambridge, Massachusetts, USA, 5-7 July 2004 (eds. S.L. Shah and J. MacGregor), 695–670. New York, NY: Published for the International Federation of Automatic Control by Elsevier Limited.

25 Talimaamar N., Babary J.P., Dochain D. (1994), Influence of the Sensor-Location on the Practical Observability of Distributed-Parameter Bioreactors, *International Conference on Control 94*, Vols **1** and **2**, 255–260.

26 Waldraff, W., Dochain, D., Bourrel, S., and Magnus, A. (1998). On the use of observability measures for sensor location in tubular reactor. *Journal of Process Control* 8: 497–505.

27 Muñoz-Tamayo R., Aceves-Lara C. A. and Bideaux C. (2014), Optimization of lipid production by oleaginous yeast in continuous culturem, *19th World Congress The International Federation of Automatic Control Cape Town, South Africa*, August 24–29.

28 Economou, C.N., Aggelis, G., Pavlou, S., and Vayenas, D.V. (2011). Modeling of single-cell oil production under nitrogen-limited and substrate inhibition conditions. *Biotechnology and Bioengineering* 108: 1049–1055.

29 Papanikolaou, S. and Aggelis, G. (2003). Modeling lipid accumulation and degradation in *Yarrowia lipolytica* cultivated on industrial fats. *Current Microbiology* 46 (6): 0398–0402.

30 John C. Polking , ODE Software for MATLAB, https://math.rice.edu/~dfield

31 Zhang, Y. and Henson, M.A. (2001). Bifurcation analysis of continuous biochemical reactor models. *Biotechnology Progress* 17: 647.

32 Karaaslanl, C.C. (2012). Bifurcation analysis and its applications. In: *Numerical Simulation - from Theory to Industry*, 3e (ed. M. Andriychuk). Intech.

33 Skogestad, S. and Postlethwaite, I. (1996). *Multivariable Feedback Control*. Wiley.

34 Bristol, E.H. (1996). On a new measure of interaction for multivariable process control. *IEEE Transactions on Automatic Control* AC-11: 133–134.

35 Ogunnaike, B.A. and Ray, W.H. (1994). *Process Dynamics, Modeling, and Control*. New York: Oxford University Press.

36 Horvat, P., Vrana Špoljarić, I., Lopar, M. et al. (2013). Mathematical modelling and process optimization of a continuous 5-stage bioreactor cascade for production of poly[-(R)-3-hydroxybutyrate] by *Cupriavidus necator*. *Bioprocess and Biosystems Engineering* 36: 1235–1250.

37 Sharma, A. and Arya, S.K. (2017). Hydrogen from algal biomass: a review of production process. *Biotechnology Reports* 15: 63–69.

38 Benedetti, M., Vecchi, V., Barera, S. et al. (2018). Biomass from microalgae: the potential of domestication towards sustainable biofactories. *Microbial Cell Factories* 17 (1): 173.

39 Park, H. and Lee, C. (2016). Theoretical calculations on the feasibility of microalgal biofuels: utilization of marine resources could help realizing the potential of microalgae. *Biotechnology Journal* 11: 1461–1470.

40 Naik, B. (2018). Volatile hydrocarbons from endophytic fungi and their efficacy in fuel production and disease control. *Egyptian Journal of Biological Pest Control* 28: 69.

8

Regulation of a Continuously Stirred Bioreactor via Modeling Error Compensation

The aim of this work is to control the nonlinear behavior of a class of continuously stirred bioreactor with regulation purposes. In the previous chapters, a linearized representation of the state space bioreactor's model is obtained via standard identification processes employing a step disturbance in the control input. The identified linear model plus an unknown modeling error term are considered and an observer-based linearized input–output controller is proposed. The observer with a polynomial structure is able to estimate the unknown modeling error of the bioreactor's representation. The proposed methodology is applied to a sulfate-reducing process where a kinetic model with a double substrate and an inhibition term, experimentally corroborated is considered as a benchmark system. A theoretical framework is presented to demonstrate the corresponding closed-loop stability of the system and numerical simulations corroborate the analytical conclusions.

8.1 Introduction

Biochemical reactors are today important process equipment in the transformation industry [1]. Bioreactors are employed in biomedical, food, fuel, and waste industries with great success. However, despite the large advance in bioprocess engineering, several operation problems remain due to the highly complex behavior of these kinds of systems. The processes design, operation analysis, optimization, and process control are important research topics to assure satisfactory performance of the bioreactor behavior [2, 3].

From the control theory point of view, there are three main trends for the design of controllers: local approaches, (semi)global approaches based on full model knowledge, and global approaches taking into account some model uncertainties. Local approaches use the linearized model of the bioreactor around the desired operating point or an identified linear model together with linear system control results [4]. Global approaches involve mainly linearizing controllers using full model knowledge for exact model linearization. The main drawback of these approaches is that they require detailed model information. Other control techniques considering model uncertainties and observer-based controllers are proposed [5].

To reach a high-quality bioreactor performance, the process controller plays an integral part in the engineering design. Therefore, the controller is generally specially formed for specific bioreactor tasks. This is related to the fact that microorganism cultivation systems

Control in Bioprocessing: Modeling, Estimation and the Use of Soft Sensors, First Edition.
Pablo Antonio López Pérez, Ricardo Aguilar López, and Ricardo Femat.
© 2020 John Wiley & Sons Ltd. Published 2020 by John Wiley & Sons Ltd.

have numerous requirements with respect to specific conditions. In the literature, several successful methods have been proposed to implement a controller for the bioreactor operating at single or multiple steady-states. For example, Kumar et al. [6] have examined input multiplicities in a continuous bioreactor. Performing an experimental analysis, mathematical modeling for the different steady-state operating regions were developed, and PI-type controllers were proposed [7]. A recurrent neural network (RNN) based modeling method for a nonlinear bioreactor operating at a single steady state has been developed and a nonlinear model predictive controller was implemented with adequate results [8]. A genetic-algorithm- (GA-) based PI controller tuning for a bioreactor operating at a stable steady state has been also considered with satisfactory performance [9]. In addition, the unstable operation in bioreactor operating has been analyzed due to the complexity of its biochemical kinetics [10–12].

Advanced process control proposals such as internal model control (IMC), sliding mode control (SMC) and model predictive control (MPC) have been presented with interesting applications to biological systems; as is well known, proportional integral derivative (PID) controllers are still widely employed in industrial control processes because of reputation for easy implementation, structural simplicity, and and its easy online tuning. With recent tuning procedures, the system's stability also satisfies chief performance criteria such as efficient disturbance rejection, measurement noise attenuation, and smooth reference tracking. Several of the PID tuning approaches designed for stable and unstable systems require numerical computations to identify the optimal controller parameters [13].

Therefore, it is necessary to propose control designs to keep the process in adequate performance conditions [14]. On the other hand, it is known that the operation of bioprocesses is a very delicate problem since one has to deal with highly nonlinear behaviors, which are generally described by poor quality model representations [15]. In order to confront these drawbacks, several authors have designed control laws that have been able to control some important state variables by employing the dilution rate of the bioreactor as handling input.

In system identification, model error modeling (MEM) is treated in, for example, Ljung et al. [16]. However, since the focus here is on default models that have biases, or other stationary errors, and aims at preserving the physical structure of the model, the MEM path is not pursued. Methods that address the issue of biased default models for estimation exist in, e.g., model augmentation using physical knowledge [17] and proportional-integral (PI) observers [18]. The methods developed in this chapter unify these ideas with the idea of estimating a minimal description of the model bias.

In standard identification problems, the error originates from two different sources: a "variance" term, due to noise affecting the data, and a "bias" term, due to system dynamics which is not captured by the estimated nominal model (often addressed also as the model error). Clearly, the nature of these two error terms is quite different: the former is generally uncorrelated with the input signal (when the data is collected in open loop), while the latter strongly depends on the estimated nominal model and on the input used in the identification experiment. The model error is not negligible in most practical situations, especially those in which the order of the nominal model must be small (a typical requirement of robust design techniques). Moreover, while prior information on measurement noise is often available, similar hypotheses on the unmodeled dynamics seem to be less realistic [19].

From the above, it is important to note the practical importance of developing a robust controller designed to reach high-performance keeping, as much as possible, simple mathematical structures to allow potential implementation in industrial process applications. In this proposal, the control design is composed of an uncertainty estimator coupled with an inverse dynamics feedback function to provide robustness against uncertain and neglected nonlinear dynamics. The control approach is based on simple step response models.

8.2 Materials and Methods

8.2.1 Bioreactor Modeling

For biological systems, the unstructured models are the simplest of all modeling philosophies used to describe the biological process. They consider the cell mass as a single chemical species and do not consider any intracellular reactions occurring within the cell. Unstructured modeling typically describes the growth phenomena based on a single limiting substrate and considers only substrate uptake, biomass growth, and product formation in the modeling framework [20].

Kinetic structures can be selected based on the experimental behavior and the best fit, based on a correlation coefficient, forming a network of mass balance equations coupled for each state variable. The structures that can be used are: modified Levenspiel model (bacterial respiration), modified Moser–Boulton model (consumption of the carbon source), modified Monod–Loung model (biofilm production), Levenspiel model–modified Haldane (removal of the metal in the biofilm and supernatant), among others reported in the literature.

8.2.2 Mathematical Model

The goal of the recently emerging interest in mathematical modeling of bioprocesses and biotechnological systems is to understand and describe quantitatively the dynamics of living cells (e.g. metabolites, enzymes, proteins). In order to attain this purpose, new experimental procedures and modeling techniques are needed to generate and analyse relevant bioprocesses data. Kinetic modeling is an important aim in bioprocess reaction engineering in order to design, monitor, operate and control bioreactors and bioprocesses.

The in silico computational tools offer a good alternative for the identification of the main regulatory pathways and genetic circuits that control the production of mainly extracellular metabolites in bioprocesses. Once the kinetic equations and the parameters have been formulated, the model can be subjected to the numerical solution via simulation step using a system of differential equations, to study its dynamic behavior and the properties of its solution in its steady state.

8.2.3 Mass Balance Modeling

In this work, the purpose is to approximate the kinetics of anaerobic sulfate-reducing bacterium (SRB) classified as *Desulfovibrio alaskensis* 6SR growth, where a double substrate is proposed with product inhibition unstructured kinetic model in accordance with the following structure which belongs to the kinetic model proposed by Keehyun and Levenspiel [21]:

$$\mu(x_2.x_3, x_4) = \mu_{max} \left(\frac{x_2}{k_{S1} + x_2} \right) \left(\frac{x_4}{k_{S2} + x_4} \right) \left(1 - \frac{x_3}{P*} \right)^n \tag{8.1}$$

The kinetic model given by Eq. (8.1) was compared with experimental data sets and a correlation coefficient of $R^2 = 0.971$ was obtained, as reported in [22]. Now, from the above kinetic model for the specific cell growth rate, the following mass balance equations are generated, considering continuous operating mode:

Biomass (x_1) mass balance:

$$\frac{dx_1}{dt} = -Dx_1 + \mu(x_2, x_3, x_4)x_1 \tag{8.2}$$

Sulfate (x_2) mass balance:

$$\frac{dx_2}{dt} = D(x_{2,in} - x_2) - \mu(x_2, x_3, x_4)x_1 Y_1 \tag{8.3}$$

Sulfide (x_3) mass balance:

$$\frac{dx_3}{dt} = -Dx_3 + \mu(x_2, x_3, x_4)x_1 Y_2 \tag{8.4}$$

Lactate (x_4) mass balance:

$$\frac{dx_4}{dt} = D(x_{4,in} - x_4) - \mu(x_2, x_3, x_4)x_1 Y_3 \tag{8.5}$$

Acetate (x_5) mass balance:

$$\frac{dx_5}{dt} = -Dx_5 + \mu(x_2, x_3, x_4)x_1 Y_4 \tag{8.6}$$

Where the parameter set is given in Table 8.1

Table 8.1 Parameters of the proposed model.

Parameters	Value	Definition	Units
μ_{max}	0.1682 ± 0.1	Maximum specific rate	h^{-1}
k_{s1}	2296 ± 10	Saturation constant for the Keehyun and Levenspiel model	$mg\,l^{-1}$
k_{s2}	1387 ± 150	Saturation constant for the Keehyun and Levenspiel model	$mg\,l^{-1}$
n	0.615 ± 0.1	Exponential term	Dimensionless
Y_1	8.1355 ± 1	Yield	Dimensionless
Y_2	1.5885 ± 0.1	Yield	Dimensionless
Y_3	13.631 ± 4	Yield	Dimensionless
Y_4	8.5139 ± 1	Yield	Dimensionless
$P*$	613.557 ± 50	Inhibition term	$mU\,mg^{-1}$

It is important to know the dynamics of the kinetic modeling of sulfate-reducing systems, mainly when working in different initial conditions due to their environmental application in metal removal.

SRB has been used to solve a number of environmental problems, e.g. removal of metals from wastewater through the production of biogenic sulfides, followed by metal precipitation. This is illustrated in the following reactions:

$$SO_4^{2-} + 8e^- + 4H_2O \rightarrow S^{2-} + 8OH^-$$

$$S^{2-} + M^{2+} \rightarrow MS \downarrow$$

$$M : Cd, Zn, Pb$$

Many experimental studies have been conducted with SRB for the removal of heavy metals from wastewater effluents.

8.3 Input–Output Identified Model

An alternative modeling approach is to use simple input/output models, which conserve the main dynamic characteristics of the process for control purposes [23]. In this work, the feedback control design is based on input–output response models. An input–output model was determined from the reaction curve process [23].

It can be seen that the step responses are smooth, almost monotonous, and convergent, such that it is reasonable to model the input–output response with a simple stable first-order model

$$G(s) = \frac{Y(s)}{U(s)} = \frac{K}{\tau s + 1} \tag{8.7}$$

Where K is the steady-state gain and τ is a process time-constant. Based on the input–output response shown in Figure 8.4, the first-order model parameters are $K = 4.857 \, (mg \, l^{-1})/(m^3 d)$ and $\tau = 170$ hours.

A state space in time domain representation of Eq. (8.7) can be presented as:

$$\tau \dot{y} + y = Ku \tag{8.8}$$

However, as is known, this linear model is only able to reproduce the system dynamic behavior in a narrow region where the system was identified, from this, if a generalization of Eq. (8.8) can be generated in order to reproduce a wide operating region, the corresponding modeling errors of Eq. (8.8) must be considered as:

$$\tau \dot{y} + y + \zeta = Ku \tag{8.9}$$

$$\dot{\zeta} = f(y, u) \tag{8.10}$$

Here, ζ represents a class of bounded uncertain modeling error terms, which are assumed to be unknown. Where the uncertain dynamic is given by Eq. (8.10); it is assumed bounded, i.e. $\|f(y, u)\| \leq \Omega$.

8.4 Control Design

The main objective of the considered control law is to regulate the dynamic behavior of the system (8.9)–(8.10). In this case, an I/O linearizing feedback approach via plant inversion is considered, firstly under ideal conditions, i.e. without uncertain or non-modeled terms the I/O linearizing feedback is able to provide exponentially and asymptotic closed-loop stability. Under this framework, the following is presented.

The control input (8.11) is a controller for the system (8.9)–(8.10):

$$u = K^{-1}(\tau \dot{y} + y + \zeta) \tag{8.11}$$

The desired closed-loop dynamic is suggested as:

$$\dot{y} = -g_1(y - y_{sp}) \tag{8.12}$$

Therefore, the final structure of the named ideal controller is:

$$u = K^{-1}(\tau g_1(y - y_{sp}) + y + \zeta) \tag{8.13}$$

Again, note that the control law (8.13) is not realizable because the term ζ is unknown.

However, when non-ideal conditions and uncertain terms are present, this control approach is not realizable. From the above, a strategy must be proposed in order to compensate the uncertain terms and reach a realizable control design, with this purpose uncertainty observer-based controllers have been presented in the open literature [24, 25].

To avoid the above drawback, the following observer for the system (8.9)–(8.10) is now considered:

$$\tau \dot{\hat{y}} + \hat{y} + \hat{\zeta} = Ku + \sum_{i=1}^{m} K_i (y - \hat{y})^{2i-1} \tag{8.14}$$

$$\dot{\hat{\zeta}} = \sum_{i=1}^{m} \overline{K}_i (y - \hat{y})^{2i-1} \tag{8.15}$$

From the above the non-ideal controller is now given by Eq. (8.16) as follows:

$$u = K^{-1}(-\tau g_1(y - y_{sp}) + y + \zeta) \tag{8.16}$$

Where $u = K^{-1}(\tau g_1(y - y_{sp} + y + \zeta)$ is provided by the observer (8.14)–(8.15).

Now, the closed-loop dynamic of the system (8.9)–(8.10) under the control is:

$$\dot{y} = -g_1(y - y_{sp}) + \tau^{-1}(\hat{\zeta} - \zeta) \tag{8.17}$$

To demonstrate the closed-loop performance of the proposed methodology and without loss of generality, assuming $y_{sp} = 0$, the Eq. (8.17) is solved as:

$$y = y_0 \exp(-g_1 t) + \exp(-g_1 t) \int_0^t \exp(g_1 \sigma) \tau^{-1} g_1^{-1}(\hat{\zeta} - \zeta) d\sigma \tag{8.18}$$

Taking norms of both sides of Eq. (8.18) yields an inequality, which is limited after boundedness assumptions:

$$0 \leq \lim_{t \to \infty} \sup \|y\| \leq \lim_{t \to \infty} \sup(\|y_0\| \exp(-g_1 t) + \exp(-g_1 \sigma) \tau^{-1} g_1^{-1} \|\hat{\zeta} - \zeta\| d\sigma) \tag{8.19}$$

By solving the above inequality applying the convergence result for the term $\|\hat{\zeta} - \zeta\| = \lim_{t \to \infty} \sup \frac{\Omega}{K_1}$, the following is obtained:

$$0 \leq \lim_{t \to \infty} \sup \|y\| \leq \lim_{t \to \infty} \sup \left(\tau^{-1} g_1^{-1} \frac{\Omega}{K_1} \right) \tag{8.20}$$

The convergence characteristics of the proposed observer are given at the end of the proof.

Consider $\omega = \Psi(x) + \Delta gu + h(x)d$ the following extended system [26, 27]:

$$\dot{x} = Ax + g_0 u + w \tag{8.21}$$

$$\dot{w} = \wp(x, \bar{u}) \tag{8.22}$$

Where $\bar{u} = (u, d)$, $\wp := \mathfrak{R}^{n+q+z} \to \mathfrak{R}^n$ is an unknown vector field which it is assumed satisfies a Lipschitz condition with respect to the vector x, i.e. and uniformly bounded with respect to u

$$\|w(x, \bar{u}) - \hat{w}(\hat{x}, \bar{u})\| \leq L\|x - \hat{x}\| \tag{8.23}$$

and considering that $|\wp| \leq \Omega < \infty$.

As mentioned above, for control purposes, an estimation of the uncertain term w is needed in order to made the considered control realizable, therefore the following uncertainty observer is considered [28–30].

Proposition 8.1 The following dynamical system is an asymptotic observer for the system (8.21)

$$\dot{\hat{x}} = A\hat{x} + g_0 u + \hat{w} + \sum_{i=1}^{m} K_i(y - C\hat{x})^{2i-1} \tag{8.24}$$

$$\dot{\hat{w}} = \sum_{i=1}^{m} \overline{K}_i(y - C\hat{x})^{2i-1} \tag{8.25}$$

Considering the following assumptions:

- w is observable with respect to $\{u,y\}$, that is to say, w satisfies a differential polynomial P in terms of $\{u, y\}$ and some of their time derivatives

$$P(w, u, \dot{u}, \ldots, y, \dot{y}, \ldots) = 0$$

- K_1 is selected from the following Ricatti algebraic equation, which has a symmetric and positive definite solution P for some $\rho > 0$.

$$(A - K_1 C)^T P + P(A - K_1 C) + L^2 PP + I + \rho I = 0 \tag{8.26}$$

- K_i is selected such that $\lambda_{\min}(PK_i C) \geq 0$, $2 \leq i \leq m$

Defining the estimation error as:

$$\xi^T = (\xi_x, \xi_w)$$

where:

$$\xi_x = x - \hat{x} \tag{8.27}$$

$$\xi_w = w - \hat{w} \tag{8.28}$$

Therefore the corresponding dynamic equation of the estimation error is:

$$\dot{\xi}_x = (A - K_1 C)\xi_x - \sum_{i=2}^{m} K_i(C\xi_x)^{2i-1} + w - \hat{w} \tag{8.29}$$

$$\dot{\xi}_w = \wp - K_1(y - C\hat{x}) - \sum_{i=2}^{m} \overline{K}_i(y - C\hat{x})^{2i-1} \tag{8.30}$$

Sketch of proof of Proposition 8.1

Now, let us to consider the following Lyapunov function candidates:

$$\Gamma = \Gamma_1 + \Gamma_2$$

$$\Gamma_1 = \xi_x^T P \xi_x \tag{8.31}$$

$$\Gamma_2 = \tfrac{1}{2}\xi_w^2 \tag{8.32}$$

Where $0 < P = P^T$

Now, $\dot{\Gamma}_1 = \dot{\xi}_x^T P \xi_x + \xi_x^T P \dot{\xi}_x$

$$\dot{\Gamma}_1 = \xi_x^T[(A - K_1 C)^T P + P(A - K_1 C)]\xi^T + 2\xi_x^T P[w - \hat{w}] - 2\sum_{i=2}^{m}(C\xi_x)^{2i-1} PK_1 C\xi_x \tag{8.33}$$

Considering the Lipschitz condition we have:

$$2\xi_x^T P[w - \hat{w}] \le L^2 \xi_x^T PP\xi_x + \xi_x^T \xi_x \tag{8.34}$$

Applying the Rayleigh inequality and considering $\lambda_{\min}(PK_iC) \ge 0$

$$-\xi_x^T PK_i C\xi_x \le -\lambda_{\min}(PK_iC)\|\xi_x\|^2 \tag{8.35}$$

Therefore, applying the Cauchy–Schwartz inequality:

$$\dot{\Gamma}_1 \le \xi[(A - K_1 C)^T P + P(A - K_1 C) + L^2 PP + \Pi]\xi^T - 2\sum_{i=2}^{m}(C\xi_x)^{2i-2}\lambda_{\min}(PK_iC)\|\xi_x\|^2 \tag{8.36}$$

$$\dot{\Gamma}_1 \le -\rho\|\xi_x\|^2 - 2\sum_{i=2}^{m}(C\xi_x)^{2i-2}\lambda_{\min}(PK_iC)\|\xi_x\|^2 \le 0 \tag{8.37}$$

then:

$$\dot{\Gamma}_1 \le -\left(\rho + 2\sum_{i=2}^{m}(C\xi_x)^{2i-2}\lambda_{\min}(PK_iC)\right)\|\xi_x\|^2 \le 0$$

Taking into account that:

$$\left(\rho + 2\sum_{i=2}^{m}(C\xi_x)^{2i-2}\lambda_{\min}(PK_iC)\right) > 0 \tag{8.38}$$

now:

$$\dot{\Gamma}_2 = \xi_w \dot{\xi}_w = \xi_w\left(\wp - K_1(y - C\hat{x}) - \sum_{i=2}^{m}K_i(y - C\hat{x})^{2i-1}\right) \tag{8.39}$$

$$\dot{\Gamma}_2 = \wp\xi_w - K_1 C\xi^2_w - \sum_{i=2}^{m}K_i(C\xi_w)^{2i} \tag{8.40}$$

Maximizing the above equation by applying the Cauchy–Schwartz inequality, the following holds:

$$\dot{\Gamma}_2 \le \Omega\,\|\xi_w\| - K_1 C\|\xi_w\|^2 - \sum_{i=2}^{m}K_i(C\xi_w)^{2i} \tag{8.41}$$

$$\dot{\Gamma}_2 \le -(K_1 C\|\xi_w\| - \Omega)\,\|\xi_w\| - \sum_{i=2}^{m}K_i(C\xi_w)^{2i} \le 0 \tag{8.42}$$

considering that: $\sum_{i=2}^{m} K_i(C\xi_w)^{2i} > 0$.

Also, K_1 can be selected to generate $K_1 C \|\xi_w\| - \Omega > 0$.

Therefore:

$$\dot{\Gamma}_2 \leq 0 \tag{8.43}$$

Note that $\dot{\Im}_2$ is negative on the set $\left\{ \|\xi_w\| \leq \lim_{t \to \infty} \sup \frac{\Omega}{K_1 C} \right\}$.

From the above, it can be concluded that:

$$\dot{\Gamma} \leq 0 \tag{8.44}$$

8.5 Main Results

In this section, the corresponding results of the proposed control method are shown. Sulfate-reducing bacteria (BSR) have a high impact on the ecological carbon and sulfur cycles, and mineralize organic matter in anaerobic environments. BSRs are widely distributed, and in environments with low sulfate levels, such as bodies of fresh water, they have relevance in the mineralization of organic matter.

An anaerobic bioreactor was simulated to evaluate the advantages of the proposed control law comparing the open-loop and the closed-loop performance. Numerical simulations were carried out employing the 23s MatlabTM R2009 library to solve ordinary differential equations, the following initial conditions for the corresponding concentrations were considered: $x_{1,0} = 100$ mg l^{-1}, $x_{2,0} = 6000$ mg l^{-1}, $x_{3,0} = 50$ mg l^{-1}, $x_{4,0} = 5950$ mg l^{-1}, and $x_{5,0} = 25$ mg l^{-1}.

The experimental data served to obtain the nine parameter values of the model developed (see Table 8.1). The parameter values were determined by nonlinear regression employing the Levenberg–Marquardt (L–M) (Table 8.2). The L–M gradient descent method converges well for problems (least squares minimization algorithm), which is a hybrid of the Gauss–Newton and the steepest descent methods.

These statistics used absolute values rather than squared differences (as in their originally specified counterparts) [31]. Interpretation of correlation-based measure 0.971 indicates that the model explains 97.1% of the variability in the observed data. With the indices of the agreement, any value (accepting 0.0 and 1.0) is difficult to interpret because of its physical meaning.

Figures 8.1 and 8.2 are related to the open-loop behavior of the mass concentrations of all the state variables, a relatively high value of the sulfate concentration is observed, around 3700 mg l^{-1} and lactate concentration, around 4000 at steady state condition, this is an indicator of the poor performance of the bioreactor for sulfate removal purposes when the dilution rate is $D = 0.025$ h^{-1}.

As the sulfide concentration can be easily measured spectrometrically, it was measured in the corresponding system to validate experimental data and dynamics in a closed loop, it is also necessary to mention that the hydrogen sulfide concentration will not be the variable to control only its internal dynamics due to its importance for the environmental systems.

Employing the information on the open-loop phase portrait it was observed that the minimum at the sulfide concentration was around 225 mg l^{-1}, and closed-loop at around 56 mg l^{-1} (Figure 8.1).

Table 8.2 Correlation coefficients

Variable	R^2
x_1	0.97
x_2	0.95
x_3	0.97
x_4	0.98
x_5	0.98
Average	0.971

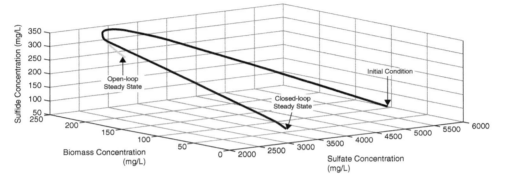

Figure 8.1 Dynamics of biomass, sulfate and sulfide concentrations.

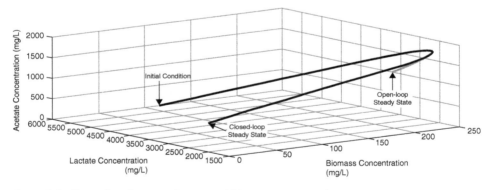

Figure 8.2 Dynamics of acetate, lactate and biomass concentrations.

The closed-loop behavior of the continuous bioreactor in Figures 8.3 and 8.4 where the proposed controller is turned-on at time = 75 hours, considering that the sulfate concentration is the measured and controlled state variable, the consuming sulfate concentration in the medium can be measured by turbid metric method based on the precipitation of barium, which is a fast and simple method [29], the control gain is selected as $k = 10 \, h^{-1}$.

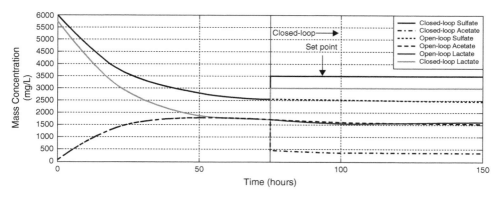

Figure 8.3 Open-loop and closed-loop dynamics of acetate, lactate, and sulfate.

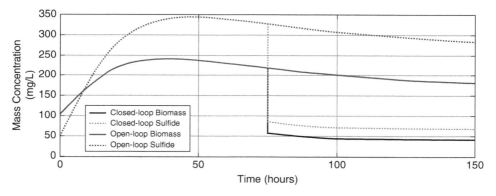

Figure 8.4 Open-loop and closed-loop dynamics of sulfide and biomass.

Firstly, the sulfate concentration is forced to reach a set point of $3500\,\mathrm{mg\,l^{-1}}$ with a setting time of the order of 30 hours without overshoots in this dynamic behavior, furthermore, to show, via numerical simulations, the capability of the proposed methodology at time $= 75$ hours, again the performance of the bioreactor is satisfactory and the new setpoint is reached without difficulties; note that all the uncontrolled mass concentrations (zero dynamics) remain with stable dynamic closed-loop response Figure 8.4.

On theother hand, Figure 8.5 shows the performance of the control effort given by the dynamic behavior of the dilution rate, the control input has an open-loop nominal value of $0.025\,\mathrm{h^{-1}}$, when the controller is turned on the dilution rate is moved to around $0.09\,\mathrm{h^{-1}}$ to reach a steady state of $0.063\,\mathrm{h^{-1}}$ in a smooth way for the considered set point.

Figures 8.1–8.5 show the corresponding performance of the uncontrolled states, also here, stable behavior was observed when the system was operated in the closed-loop mode.

It is important to note that the required effort of the proposed controller belongs to a physically realizable domain. Finally, a comparison with a class of PI controller under the same control gain was done via a performance index named integral time-weighted square error (ITSE) which more penalizes large control errors at long times, Figure 8.6 shows the better performance of the proposed methodology [32]. This result is due to the ability of the controller to eliminate offset properties, which is not presented by the other PI controller.

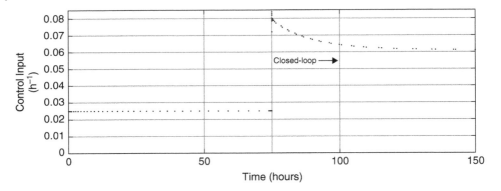

Figure 8.5 Control effort.

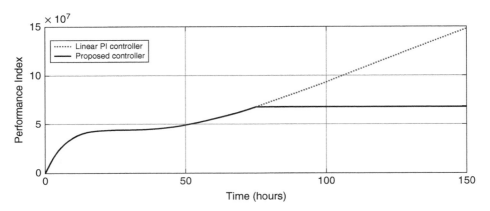

Figure 8.6 Performance index.

High sulfate containing wastewaters are generated from various industrial activities. These include pulp and paper industries, mining and mineral processing, production of explosives, scrubbing of flue gases, food processing, and petrochemical industries. This methodology can be used to evaluate the behavior of the sulfate-containing wastewater and to determine the properties of the dynamic system such as the existence of an equilibrium point, the uniqueness or the multiplicity of equilibrium points, local or global asymptotic stability of equilibrium points [33]. Therefore, the approach presented here may be considered to be a valid method to design controllers for this class of systems [34–36].

8.6 Concluding Remarks

In this work we present a kinetic model experimentally corroborated for a sulfate-reducing process, this model is employed as a benchmark on a continuously stirred tank bioreactor, where both an unknown modeling error term is considered and an observer-based linearized input–output controller. In order to regulate the sulfate concentration in the bioreactor sulfate concentration measurements manipulating the dilution rate (input flow).

The proposed controller is able to lead to the sulfate concentration to the required set points, with a satisfactory effort as shown by the numerical simulations, in agreement with the closed-loop stability analysis. For comparison purposes, a PI controller is implemented and the proposed controller shows better performance.

List of Figures

List of Tables

References

1 Santos-Sánchez, N.F., Salas-Coronado, R., Santos-Sánchez, O.J. et al. (2016). On the effects of the temperature control at the performance of a dehydration process: energy optimization and nutrients retention. *International Journal of Advanced Manufacturing Technology* 86: 3157.

2 Zweigerdt, R. (2009). Large scale production of stem cells and their derivatives. *Advances in Biochemical Engineering/Biotechnology* 114: 201–235.

3 Baños-Rodríguez, U., Santos, O., Beltrán, R.I., and Vázquez-Rodríguez, G.A. (2012). Model-based control of a fed-batch biodegradation process by the control Lyapunov function approach. *Chemical Engineering Journal* 189–190: 256–263.

4 López-Pérez, P.A., Neria-González, M.I., and Aguilar-López, R. (2012). Cadmium concentration stabilization in a class of continuous sulfate reducing bioreactor via sulfide concentration control. *Chemical Papers* 67: 326–335.

5 Saadatjoo, F., Derhami, V., and Karbassi, S.M. (2011). Stabilization of simultaneous linear multivariable systems while improving time-response using genetic algorithms. *International Journal of Innovative Computing, Information, and Control* 7: 151–160.

6 Giriraj Kumar, S.M., Jain, R., Anantharaman, N. et al. (2008). Genetic algorithm based PID controller tuning for a model bioreactor. *Indian Institute of Chemical Engineers* 50 (3): 214–226.

7 Kumar, S.V.S., Kumar, V.R., and Reddy, G.P. (2005). Nonlinear control of bioreactors with input multiplicities—an experimental work. *Bioprocess and Biosystems Engineering* 28 (1): 45–53.

8 Sivakumaran, N., Radhakrishnan, T.K., and Babu, J.S.C. (2006). Identification and control of bioreactor using recurrent networks. *Instrumentation Science and Technology* 34 (6): 635–651.

9 Latha, K., Rajinikanth, V., and Surekha, P.M. (2013). PSO-based PID controller design for a class of stable and unstable systems. *ISRN Artificial Intelligence* 11: 543–607.

10 Pramod, S. and Chidambaram, M. (2000). Closed-loop identification of transfer function model for unstable bioreactors for tuning PID controllers. *Bioprocess Engineering* 22 (2): 185–188.

11 Galluzzo, M. and Cirino, C. (2013). Sliding mode fuzzy logic control of an unstable bioreactor. *Chemical Engineering Transactions* 32.

12 Olivieria, G., Russob, M.E., Mancusic, E. et al. (2013). Non-linear analysis of feedback controlled aerobic cultures. *Chemical Engineering Transactions*: 32.

13 Rodríguez-Guerrero, L., López, O., and Santos, O. (2012). Object-oriented optimal control for a batch dryer process. *International Journal of Advanced Manufacturing Technology* 58 (1–4): 293–307.

14 Wang, Y., Fang, X., An, F. et al. (2011). Improvement of antibiotic activity of *Xenorhabdus bovienii* by medium optimization using response surface methodology. *Microbial Cell Factories* 14, 10: 98.

15 Bernard, O., Hadj-Sadok, Z., and Dochain, D. (2000). Software sensors to monitor the dynamics of microbial communities: application to anaerobic digestion. *Acta Biotheoretica* 48 (3–4): 197–205.

16 Ljung, L. (1999). Model validation and model error modeling. In: *Proc. of the °Aström Symposium on Control* (eds. B. Wittenmark and A. Rantzer), 15–42, Lund, Sweden, Aug. Studentliteratur, Lund, Sweden.

17 Andersson, P., and Eriksson, L. (2004), Cylinder air charge estimator in turbocharged SI engines. In *Electronic Engine Controls*, 2004-01-1366 in *SAE Technical paper series* SP-1822.

18 Koenig, D. and Mammar, S. (2002). Design of proportional-integral observer for unknown input descriptor systems. *IEEE Transactions on Automatic Control* 47 (12): 2057–2062.

19 Michel Gevers, *IDENTIFICATION FOR CONTROL: ACHIEVEMENTS AND OPEN PROBLEMS 1*, http://perso.uclouvain.be/michel.gevers/PublisMig/C128.pdf

20 Schü Gerl, K. and Bellgardt, H. (2000). *Bioreaction Engineering*. Berlin, Heidelberg: Springer.

21 Han, K. and Levenspiel, O. (1987). Extended Monod kinetic for substrate, product, and cell inhibition. *Biotechnology and Bioengineering* 32: 430–437.

22 Peña-Caballero, V. (2013), Analysis of the operation of hybrid processes for the removal of Cr(VI), Ph.D. Thesis. CINVESTAV-IPN, México, in spanish.

23 Ogunnaike, B.A. and Ray, W.H. (1994). *Process Dynamics, Modeling, and Control*. New York: Oxford University Press.

24 Karafyllis, I. and Kravaris, C. (2005). Robust output feedback stabilization and nonlinear observer design. *Systems & Control Letters* 54 (10): 925–938.

25 Aguilar, R., González, J., Barrón, M.A. et al. (2001). Robust PI^2 controller for continuous bioreactors. *Process Biochemistry* 36 (10): 1007–1014.

26 Aguilar-López, R., Gonzalez-Trejo, J., and Alvarez-Ramirez, J. (1997). Temperature regulation of a class of continuous chemical reactor based on a nonlinear Luenberger-like Observer. *Journal of Chemical Technology & Biotechnology* 70 (3): 209–216.

27 Gonzalez, J., Fernandez, G., Aguilar, R., and Alvarez-Ramirez, J. (2001). Sliding mode observer-based control for a class of bioreactors. *The Chemical Engineering Journal* 83 (1): 25–32.

28 Mata-Machuca, L., Martínez-Guerra, R., and Aguilar-López, R. (2011). Differential algebraic estimator for the monitoring of a class of partially known bioreactor models. *Revista mexicana de ingeniería química* 10 (2): 313–320.

29 Khalil, H.K. (1996). *Nonlinear Systems*, 2e. NJ, U.S.A: Prentice Hall.

30 López-Pérez, P.A., Neria-González, I., and Aguilar-López, R. (2015). A biotechnological alternative in the cadmium removal at high concentration: *Desulfovibrio alaskensis* 6sr. *International journal of Environmental Science and Technology* 12 (6): 1975–1986.

31 Marquardt, D.W. (1963). An algorithm for least-squares estimation of nonlinear parameters. *Journal of the Society for Industrial and Applied Mathematics* 11 (2): 431–441.

32 Seborg, D.E., Edgar, T.F., and Mellichamp, D.A. (1989). *Process Dynamics and Control*. New York: Wiley.

33 Moreno, J.A. and Vande Wouwer, A. (2014). Special issue on nonlinear modeling, estimation, and control of biological systems. *Bioprocess and Biosystems Engineering* 37: 1–3.

34 Meraz, M., Ibarra-Valdez, C., and Alvarez-Ramirez, J. (2016). Modeling-error compensation approach for extremum-seeking control of continuous stirred tank bioreactors with unknown growth kinetics. *Industrial & Engineering Chemistry Research* 5514: 4071–4079.

35 Malisoff, M. and Zhang, F. (2013). Adaptive control for planar curve tracking under controller uncertainty. *Automatica* 49 (5): 1411–1418.

36 Penttonen, J., Lehtonen, M., and Muhammad, S. (2017). A novel 4-limb earth fault compensation reactor for distribution network. *International Transactions on Electrical Energy Systems* 27, e2433.

9

Development of Virtual Sensor Based on the Just-In-Time Model for Monitoring of Biological Control Systems

Real-time monitoring of physiological characteristics during a cultivation process is of great importance in bioprocesses. Biological control involves the use of beneficial organisms for metabolite production that reduces the negative effects of plant pathogen disease suppression. However, whilst experimentation has elucidated the physiological basis of a number of host–pathogen–antagonist interactions, mathematical models are less well developed and the performance of biological control is beset by variability. This chapter addresses a novel virtual sensor for online monitoring of biological control of *Trichoderma harzianum* against *Cladosporium cladosporioides* isolated from cocoa. A mathematical model, validated experimentally, was used as a virtual plant together with the concentration of *T. harzianum* for the real-time estimation of the concentrations of β-1,3 glucanases, chitinases, lipases, proteases, proteolytic enzymes that according to their degree of induction could participate in the antagonistic effect of *T. harzianum* VSL291 against *C. cladosporioides*, on the other hand, the concentration that cannot be estimated was the chitinase. A theoretical analysis is presented to provide a sketch of proof to demonstrate the convergence properties of the proposed observer. Numerical simulations are provided to show the effectiveness of the proposed observer where a comparison with an extended Luenberger observer is carried out.

9.1 Introduction

Cacao (*Theobroma cacao* L.) is a crop of great commercial value in many tropical regions of the world, whose fruit is processed for use in various products. The most important diseases in cocoa are moniliasis [1], witch's broom [2] and black spot [3]. One of them is a scab, also known as *Cladosporium* rot, the etiology of which is still under study [4]. This disease can infest and kill seedlings and branches, delay production and flowering, and affect the quality of the fruits, and although the fungus does not cause deterioration of the pulp, but the appearance and acceptance of the fruits in the market is reduced [5]. *Cladosporium* is a genus of fungi that has been isolated worldwide which includes saprophytic, phytopathogenic, and species that are pathogenic to humans [6]. *Cladosporium herbarum* and *C. cladosporioides* are among the most prevalent species of internal and external environments according to the more than 772 species that comprise this genus and the 180 strains registered with the data bank International Universal Protein Resource [7].

Control in Bioprocessing: Modeling, Estimation and the Use of Soft Sensors, First Edition.
Pablo Antonio López Pérez, Ricardo Aguilar López, and Ricardo Femat.
© 2020 John Wiley & Sons Ltd. Published 2020 by John Wiley & Sons Ltd.

On the other hand, the antagonistic capacity of *Trichoderma* is very variable, the myco-parasitic that presents is a complex process that includes a series of successive events. In addition, the interaction of *Trichoderma* with its host has been shown to be specific and controlled by lectins present in its cell wall. Alternatively, observations under the microscope have suggested that *Trichoderma* produces and secretes mycolytic enzymes responsible for the partial degradation of the cell wall of its host [8].

Several authors suggest that mycoparasitism is the main mechanism of action of *Trichoderma*, through which it penetrates the cells of the pathogen causing extensive damage such as degradation of the cell wall, retraction of the plasmatic membrane of the wall and lastly the disorganization of the cytoplasm [9]. In particular, for biochemical processes, the optimal performances depend on available information. Because these variables are frequently associated with the process output quality, they are very important for process control and monitoring.

The principal obstacle in the control industry process is to measure the input variable and the output variable in order to drive this output into the set point of the process. Several high price instruments have been designed for this purpose.

For this reason, it is of great importance to deliver additional information about process variables, which is the role of the software sensor [10]. Nevertheless, virtual sensors have several advantages besides their lower cost. For example, virtual sensors could be more easily added as retrofits in a number of important applications, such as measurement of biomass or metabolites. There has been a rapid development of virtual sensing technology over the past decade within a number of different domains, including avionics, autonomous robots, telemedicine, traffic, automotive, nature and building monitoring, and control. However, the development and especially the implementation of advanced monitoring and control strategies on real bioprocesses are difficult because of an absence of reliable instrumentation for the biological state variables, i.e. the substrates, biomass, and product concentrations. For example, the quality of monitored data, precision data, time delay, and frequency of sampling required are functions of the accuracy of bio-sensors, and usually require more sophisticated measurement devices, which can have several drawbacks, e.g. sterilization, discrete-time (and often rare) samples, relatively long processing (analysis) time. In many cases the state variables, are not on-line (and real-time), this is related to high-cost sensors and extreme operating conditions, these facts, together with the nonlinearity and parameter uncertainty of the bioprocesses requires an enhanced modeling effort and modern state estimation and identification strategies [11].

The measurement characteristic category refers to whether the desired virtual sensor outputs are transient or steady-state variables. A transient virtual sensor incorporates a transient model to predict the transient behavior of an unmeasured variable in response to measured transient inputs.

A solution to these latter problems can be found through the design of software sensors, this estimation approach is based on combining some available measurement devices to provide signals such as dissolved oxygen, glucose, pH and temperature and a mathematical model, in order to provide continuous time estimates of non-measured variables on-line, the estimation algorithm is called a state observer [12].

Kadlec et al. [13] defined a software sensor as a combination of the words "software," because the models are generally computer programs (algorithm), and "sensors," because the models are delivering similar information as their hardware equivalents. Virtual sensors provide indirect measurements of abstract conditions (that, by themselves, are not physically measurable) by combining sensed data from a group of heterogeneous physical sensors.

Signals from these individual sensors can be used in calculations within a virtual sensor to determine if, for example, a crane has exceeded its safe working load. Another benefit of the virtual sensor is that it can be used to mask the explicit data sources (sensors) that provide data.

The first class of observers including classical observers like Luenberger and Kalman observers, and nonlinear observers are based on perfect knowledge of the model structure [14–16].

Due to the above, this project addresses a novel virtual sensor for online monitoring of biological control of *T. harzianum* against *C. cladosporioides* using a mathematical model validated experimentally which was used as a virtual plant together with the concentration of *T. harzianum* for the real-time estimation of the concentrations of β-1,3 glucanases, chitinases, lipases, proteases, proteolytic enzymes that according to their degree of induction could participate in the antagonistic effect of *T. harzianum* VSL291 against *cladosporium*. Numerical simulations are provided to show the effectiveness of the proposed observer where a comparison with a standard observer is done.

9.2 Materials and Methods

9.2.1 Kinetic and Simulated Mycoparasitism *T. harzianum – C. cladosporioides*

The enzymes produced by *T. harzianum* degraded the cell walls of *C. cladosporioides* after 50 hours [17]. In this chapter a kinetic model was proposed that interprets and represents the mechanism of biological control through the growth rate of fungi and the secretion of proteins. The model was based on the balance of mass via unstructured kinetics represented by differential equations as a function of time (dynamic model). For the solution of the model, a graphical programming network was designed representing the reaction mechanism of mycoparasitism as a function of the enzymes involved. For the numerical solution, the Runge–Kutta method was used.

For the confrontation experiment, the *T. harzianum* VSL291 strain against *C. cladosporioides* strains was used as an antagonist, according to the technique proposed by Szekeres et al. [17]. In order to determine the simulated mycoparasitism of *T. harzianum* on phytopathogenic fungus, the cell walls of *C. cladosporioides* were used as the only source of carbon. In 50 ml flasks 50 ml of minimal medium was placed (containing: $MgSO_4 \cdot 7H_2O$, 0.24 g, KCl, 0.24 g, NH_4NO_3, 1.2 g, $ZnSO_4 \cdot 7H_2O$, 0.0024 g $\cdot$ $MgCl_2 \cdot 7H_2O$, 0.0024, K_2HPO_4, 1.08 g, $FeSO_4 \cdot 7H_2O$, 0.0024 g, pH 5.5, to 1200 ml) and $10\,g\,l^{-1}$ cell wall of the milled mycelium of *C. cladosporioides*. Each flask was seeded with a mycelial from *T. harzianum* strain VSL291 and fermented for three days. The enzyme activity was determined for β-1,3-glucanase [18], xylanase [19], chitinase [20], and protease [21]. The quantification of reducing sugars was performed by the method described by Miller (1959). A more detailed explanation of the experimental part can be seen in Romero et al. [22].

9.2.2 Mathematical Model

The goal of the recently emerging interest in mathematical modeling of bioprocesses and biotechnological systems is to understand and describe quantitatively the dynamics of living cells (e.g. metabolites, enzymes, proteins). In order to attain this purpose, new experimental procedures and modeling techniques are needed to generate and analyse relevant bioprocesses data. Kinetic modeling is an important aim in bioprocess reaction engineering in order to design, monitor, operate and control bioreactors and bioprocesses.

A state observer is a device that estimates the unknown states of a dynamical system. It utilizes the system model and measurements of the system inputs and outputs for the estimation process. The proposed kinetic model predicted concentrations of chitinases, β-1,3 glucanase enzymes, proteases, lipases, xylanases produced by *T. harzianum* cell walls of *C. cladosporioides* [23]. Balances of the model are presented as follows EDOs (Table 9.1):

T. harzianum (*S*) mass balance:

$$\frac{dS}{dt} = \frac{\mu_{max}}{Y} * \left(1 - \frac{S}{k_s}\right)^{\alpha} * \left[\frac{X}{k_x + X}\right] \tag{9.1}$$

C. cladosporioides (*X*) mass balance:

$$\frac{dX}{dt} = \mu_{max} * \left(1 - \frac{S}{k_s}\right)^{\alpha} * \left[\frac{X}{k_x + X}\right] * B * \left[\frac{\varphi_1}{\varphi_1 + Q}\right]$$
$$* \left[\frac{\varphi_2}{\varphi_2 + P}\right] * \left[\frac{L}{\varphi_3 + L}\right] * \left(\frac{1}{G^{\varphi_4}}\right) \tag{9.2}$$

Xylanases (*B*) mass balance:

$$\frac{dB}{dt} = \left[\lambda_B * \left(1 - \frac{B}{k_B}\right)^{\beta} - k_d * X * B + \lambda_1 * \exp^{(-\lambda_2 * X * t)} - \lambda_1 * \exp^{(-\lambda_3 * X * t)}\right] * S \tag{9.3}$$

Proteases (*P*) mass balance:

$$\frac{dP}{dt} = \left[\rho_P * \left(1 - \frac{P}{k_P}\right)^{\chi} - k_d * X * P + \rho_1 * \exp^{(-\rho_2 * X * t)}\right] * S \tag{9.4}$$

Chitinases (*Q*) mass balance:

$$\frac{dQ}{dt} = \left[\kappa_Q * \left(1 - \frac{Q}{k_Q}\right)^{\delta} - k_d * X * Q + \kappa_1 * \exp^{(-\kappa_2 * X * t)}\right] * S \tag{9.5}$$

β-1,3 glucanases (*G*) mass balance:

$$\frac{dG}{dt} = \left[\nu_G * \left(1 - \frac{G}{k_G}\right)^{\varepsilon} - k_d * X * G + \nu_1 * \exp^{(-\nu_2 * X * t)}\right] * S \tag{9.6}$$

Lipases (*L*) mass balance:

$$\frac{dL}{dt} = \left[\sigma_L * \left(1 - \frac{L}{k_L}\right)^{\eta} - k_d * X * L + \sigma_1 * \exp^{(-\sigma_2 * X * t)}\right] * S \tag{9.7}$$

Moreover, assessing the effect of operational factors via parameter values may provide useful information for improving design configuration and the operation of bioreactors. It is worth mentioning that many of these constants are in the range with some experiments related to mycoparasitism [24–26].

Table 9.1 Parameters of model proposed.

Parameters	Value	Definition	Units
$\mu_{max} = 0,0156$	0.014 ± 0.01	Maximum specific rate	$[cm^2\,h^{-1}]$
$k_s = 0,95$	0.9 ± 0.01	Inhibition term for the Luong model	Dimensionless
$k_x = 15$	14 ± 1.9	Saturation constant for *T. harzianum*	$mU\,mg^{-1}$
$\alpha = 1,6$	1.6 ± 0.1	The exponential term for the Luong model	Dimensionless
$Y = 0,8$	0.85 ± 0.01	Yield	Dimensionless
$\varphi_1 = 2,3$	2.0 ± 0.91	Bolton model constant for chitinases over *cladosporioides*	$mU\,mg^{-1}$
$\varphi_2 = 3,9$	3 ± 0.9	Bolton model constant for proteases on *C. cladosporioides*	$mU\,mg^{-1}$
$\varphi_3 = 0,01$	0.01 ± 0.015	Bolton model constant for lipases on *C. cladosporioides*	$mU\,mg^{-1}$
$\varphi_4 = 0,01$	0.01 ± 0.015	Bolton model constant for β-1,3 glucanases on *C. cladosporioides*	$mU\,mg^{-1}$
$\sigma_L = 0,001$	$0.001 \pm 0.00.5$	Specific product speed of lipases	$mU\,mg^{-1}\,h^{-1}\,cm^{-2}$
$k_L = 15$	15 ± 1	Inhibition term for the Luong model for lipases	$mU\,mg^{-1}$
$\varepsilon = 1,6$	1.6 ± 0.5	The exponential term for the Luong model	Dimensionless
$\sigma_1 = 0,63$	0.63 ± 0.25	Constant referred to the metabolism of lipases	$mU\,mg^{-1}\,h^{-1}\,cm^{-2}$
$\sigma_2 = 1,4$	1.5 ± 0.35	Exponential constant of lipases activity	Dimensionless
$\lambda_B = 0,015$	0.010 ± 0.025	Specific product rate of xylanases	$mU\,mg^{-1}\,h^{-1}\,cm^{-2}$
$k_s = 13$	13 ± 5	Inhibition term for the Luong model for xylanases	$mU\,mg^{-1}$
$\beta = 3,12$	0.63 ± 0.25	The exponential term for the Luong model	Dimensionless
$k_d = 0.01$	0.010 ± 0.025	Constant referred to the metabolism *C. cladosporioides*	$h^{-1}\,cm^{-2}$
$\lambda_1 = 3,3$	3 ± 0.25	Constant referred to the metabolism of xylanases	$mU\,mg^{-1}\,h^{-1}\,cm^{-2}$

Table 9.1 (Continued)

Parameters	Value	Definition	Units
$\lambda_2 = 1,36$	1.5 ± 0.25	Exponential constant of xylanase activity (active)	Dimensionless
$\lambda_3 = 2,0$	2.5 ± 0.5	Exponential constant of xylanase activity (inactive)	Dimensionless
$\rho_P = 0,024$	0.045 ± 0.25	Specific product rate of proteases	$mU\,mg^{-1}\,h^{-1}\,cm^{-2}$
$k_P = 16$	15 ± 2	Inhibition term for the Luong model for proteases	$mU\,mg^{-1}$
$\chi = 1,6$	1.5 ± 0.3	The exponential term for the Luong mode	Dimensionless
$\rho_1 = 0,2$	0.5 ± 0.2	Constant referred to the metabolism of proteases	$mU\,mg^{-1}\,h^{-1}\,cm^{-2}$
$\rho_2 = 1,5$	1.5 ± 0.25	Exponential constant of protease activity	Dimensionless
$\kappa_Q = 0,0056$	0.005 ± 0.002	Specific product rate of chitinases	$mU\,mg^{-1}\,h^{-1}\,cm^{-2}$
$k_Q = 15$	11 ± 3	Inhibition term for the Luong model for quitinases	$mU\,mg^{-1}$
$\delta = 1,6$	1.5 ± 0.5	The exponential term for the Luong model	Dimensionless
$\kappa_1 = 1,2$	1.5 ± 0.4	Constant referred to the metabolism of chitinases	$mU\,mg^{-1}\,h^{-1}\,cm^{-2}$
$\kappa_2 = 3,5$	4.5 ± 0.25	The exponential constant of the activity of chitinases	Dimensionless
$\nu_G = 0,048$	0.05 ± 0.2	Specific product rate of glucanases	$mU\,mg^{-1}\,h^{-1}\,cm^{-2}$
$k_G = 15$	15 ± 3	Inhibition term for the Luong model for glucanases	$mU\,mg^{-1}$
$\epsilon = 1,6$	1.6 ± 0.5	The exponential term for the Luong model	Dimensionless
$\nu_1 = 1,25$	1.5 ± 0.4	Constant referred to the metabolism of glucanases	$mU\,mg^{-1}\,h^{-1}\,cm^{-2}$
$\nu_2 = 1,4$	1.5 ± 0.7	Exponential constant of glucanase activity	Dimensionless

9.3   On-line Monitoring (Proposed Nonlinear Observer)

The state observer provides indirect measurements (which, by themselves, are not physically measurable) in real conditions by combining sensed data from a group of heterogeneous physical sensors (in real time). Signals from individual sensors can be used in calculations within an estimator of a reconstruction state variable; in addition, it is inevitable that some deviation occurs between the trajectories of a real system and the predictions of a mathematical model.

In Figure 9.1, the output measurements are used to provide feedback to the virtual-sensor from the real process. The algorithm of the nonlinear observer uses these signals as the online available measurement data, noting that the concentration of the *C. cladosporioides* is considered as the measured system output and this allows the reconstruction of some state variables.

The concentration was quantified and collaborated experimentally through the radial growth of the fungus through the acquisition of a photographic image with a normal digital camera using a Matlab algorithm for signal processing and feeding back this information to the virtual sensor to predict concentrations that cannot be measured online with a focus and implementation of just-in-time (JIT) model based adaptive soft sensor [27].

Note: An observer comprises a real-time simulation of the system or plant, driven by the same input of the system, and by a correction term derived from the difference between the actual output of the system and the predicted output derived from the observer. This

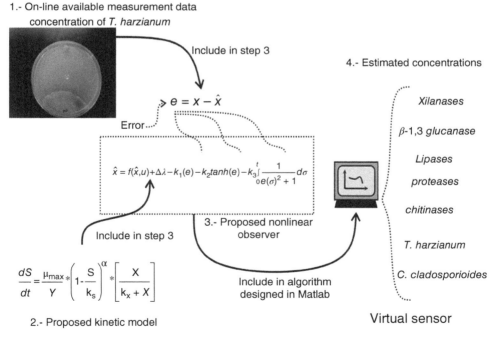

1.- On-line available measurement data
concentration of *T. harzianum*

Include in step 3

4.- Estimated concentrations

$e = x - \hat{x}$

Error

Xilanases

β-1,3 glucanase

$$\hat{x} = f(\hat{x},u)+\Delta\lambda - k_1(e) - k_2 tanh(e) - k_3 \int_0^t \frac{1}{e(\sigma)^2 + 1} d\sigma$$

Lipases

proteases

chitinases

Include in step 3

3.- Proposed nonlinear observer

T. harzianum

$$\frac{dS}{dt} = \frac{\mu_{max}}{Y} * \left(1 - \frac{S}{k_s}\right)^\alpha * \left[\frac{X}{k_x + X}\right]$$

Include in algorithm designed in Matlab

C. cladosporioides

2.- Proposed kinetic model

Virtual sensor

Figure 9.1   Virtual sensor based on JIT model for monitoring of biological control systems.

proposal [13] presents an estimation technique for systems subject to modeling errors df (parameter uncertainties), which is a realistic process situation. The modeling errors in the design of nonlinear observers can cause poor estimation performance. Therefore, an alternative state observer is introduced, which is able to converge to the real states with satisfactory performance. More explicitly, we pursue the design of observers that are sensitive to the effect of modeling error. For this purpose, proposed the following observer's structure and its corresponding convergence analysis is proposed.

Let the general structure of state observers be derived. Consider the following nonlinear system representation:

$$\dot{x} = f(x, u) \tag{9.8}$$

$$y = h(x) = Cx \tag{9.9}$$

where $x \in \mathfrak{R}^n$ the vector of the state variables; $u \in \mathfrak{R}^q$ is the control input vector; $f(\circ) : \mathfrak{R}^{n+q} \to \mathfrak{R}^n$ is a nonlinear smooth vector function and Lipschitz in x and uniformly bounded in u, $\Delta\eta$ is the modeling error and $y \in \mathfrak{R}^m$ is the vector of measured states.

The general structure of the state observer for the system (9.8–9.9) is:

$$\hat{x} = \hat{f}(\hat{x}, u, y) \tag{9.10}$$

Usually it is required that at least $u \in \mathfrak{R}^q$
$\|\hat{x}\text{-}x\| \to 0, \, as \, t \to \infty \|\hat{x} - x\| \to 0, \, ast \to \infty$ [25].

Proposition 9.1 The following dynamic system is an observer described for Eq. (9.9):

$$\hat{x} = f(\hat{x}, u) + \Delta\lambda - k_1(e) - k_2 \tanh(e) - k_3 \int_0^t \frac{1}{e(\sigma)^2 + 1} d\sigma \tag{9.11}$$

where

$$e = x - \hat{x} \tag{9.12}$$

9.3.1 Sketch of Proof of Proposition 9.1

In practice, there is an estimation error which is defined by the difference between the estimated state and the true state, considering Eqs. (9.8) and (9.11) the following equation was obtained for the observation error dynamic:

$$\dot{e} = \dot{x} - \hat{x} \tag{9.13}$$

$$\dot{e} = f(x, u) - f(\hat{x}, u) + \Delta\lambda - k_1(e) - k_2 \tanh(e) - k_3 \int_0^t \frac{1}{e(\sigma)^2 + 1} d\sigma \tag{9.14}$$

where:

$$x = [S, X, B, P, Q, G, L]^T \in \mathfrak{R}^7_+ \tag{9.15}$$

$$C = [0, 1, 0, 0, 0, 0, 0]^T$$

$\Delta \lambda$ is the modeling error

$$
f(x,u) = \begin{bmatrix}
\begin{bmatrix} \frac{\mu_{max}}{Y} * \left(1 - \frac{S}{k_s}\right)^\alpha * \left[\frac{X}{k_x + X}\right] \\ \mu_{max} * \left(1 - \frac{S}{k_s}\right)^\alpha * \left[\frac{X}{k_x + X}\right] * B * \left[\frac{\varphi_1}{\varphi_1 + Q}\right] * \left[\frac{\varphi_2}{\varphi_2 + P}\right] * \left[\frac{L}{\varphi_3 + L}\right] * \left(\frac{1}{G^{\varphi_4}}\right) \end{bmatrix} * S \\
\left[\lambda_B * \left(1 - \frac{B}{k_B}\right)^\beta - k_d * X * B + \lambda_1 * \exp^{(-\lambda_2 * X * t)} - \lambda_1 * \exp^{(-\lambda_3 * X * t)}\right] * S \\
\left[\rho_P * \left(1 - \frac{P}{k_P}\right)^\chi - k_d * X * P + \rho_1 * \exp^{(-\rho_2 * X * t)}\right] * S \\
\left[\kappa_Q * \left(1 - \frac{Q}{k_Q}\right)^\delta - k_d * X * Q + \kappa_1 * \exp^{(-\kappa_2 * X * t)}\right] * S \\
\left[v_G * \left(1 - \frac{G}{k_G}\right)^\epsilon - k_d * X * G + v_1 * \exp^{(-v_2 * X * t)}\right] * S \\
\left[\sigma_L * \left(1 - \frac{L}{k_L}\right)^\eta - k_d * X * L + \sigma_1 * \exp^{(-\sigma_2 * X * t)}\right] * S
\end{bmatrix}
$$

Taking the norm to maximize Eq. (9.14):

$$
|\dot{e}| \le |f(x,u) - f(\hat{x},u)| + \Delta\lambda - k_1|e| - k_2|\tanh(e)| - k_3 \int_0^t \frac{1}{e(\sigma)^2 + 1} d\sigma \tag{9.16}
$$

Now, taking into account the following assumptions and by the properties of the corresponding functions:

A1.

$$
|f(x,u) - f(\hat{x},u)| \le \ell|x - \hat{x}| \tag{9.17}
$$

The Lipschitz constant is $\ell > 0$

A2.

$$
|\Delta\lambda| \le T \tag{9.18}
$$

A3.

$$
|\tanh(e)| \le 1 \tag{9.19}
$$

A4.

$$
\int_0^t \frac{1}{e(\sigma)^2 + 1} d\sigma \le 1 \text{ (Property of the arctan function)} \tag{9.20}
$$

Therefore Eq. (9.16) can be expressed as:

$$
|\dot{e}| \le (\ell - k_1)|e| + [T - (k_2 + k_3)] \tag{9.21}
$$

By solving the above differential inequality:

$$
|e| \le e_0 \exp((\ell - k_1)t) + \frac{[T - (k_2 + k_3)]}{\ell}(1 - \exp((\ell - k_1)t)) \tag{9.22}
$$

Considering the matrix $L - k_1$ as a Hurwitz stable matrix

For $t \to \infty$

$$
|e| \le \frac{[T - (k_2 + k_3)]}{\ell} \tag{9.23}
$$

Remarks Note that the proportional term of the observer structure provides, as usual, stability to the observer, the observer's gain k_1 acts as a convergence rate parameter to lead to the estimation error to the closed-ball with a radius proportional to $(\Gamma - (k_2 + k_3))$, moreover the estimation error can be made as small as desired if $(k_2 + k_3) \approx \Gamma k_2 \sim \Upsilon$.

9.4 Such Approaches, Known as Proposed Just-in-Time Modeling "Hybrid Systems"

Currently, the investigations directed at the so-called data-driven models are becoming more and more frequently used. These models rely upon the methods of different advanced techniques in computational intelligence, machine learning, and mathematical modeling, and thus assume the presence of a considerable amount of data describing the modeled system's chemical, biological, and physics (i.e. signals of products, metabolites, substrates, strains) [27, 28].

Advantages of using JIT :

- bioprocess nonlinearity [29]
- changes in operation process [30]
- recursive techniques [31]
- dataset around a query data sample
- JIT model is built online [32]
- JIT model can trace the current state of the process [27]

Data-driven virtual sensors (software sensor) have gained impact due to the availability of the recorded historical dynamic in the plant data or on line. The implementations of soft sensors, however, involve some practical difficulties. To get a soft sensor automatically updated, different methods have been introduced:

- Kalman filter
- moving window techniques
- recursive methods
- ensemble approach [27].

However, these methods have some drawbacks which motivate the development and implementation of JIT model based adaptive soft sensor [29]. JIT models have been applied as soft sensors in various units and processes.

Moreover, the modeling of multiple input/output systems with black-box tools usually requires large amounts of data for reliable predictions to be obtained. In view of the above reasons, approaches that combine the properties of mechanistic (white-box) models with those of empirical (black-box) techniques, integrating the best of both paradigms, seem to be quite appealing. Such approaches, known as "hybrid models," aim to achieve good extrapolation properties, some degree of process behavior rationalization, facility of model development, and focus on relevant phenomenological parameter fitting.

Remarks   A JIT model based soft sensor has been outlined to provide a pathway toward developing improved and more practical adaptive soft sensors based on JIT models [27]. In this work, only the input ("y" in real time) of the database is related to feed the soft sensor (observer) leaving for future work the hybrid model with JIT to feed the soft sensor via the update of the model or directly to a controller as shown in Figure 9.2.

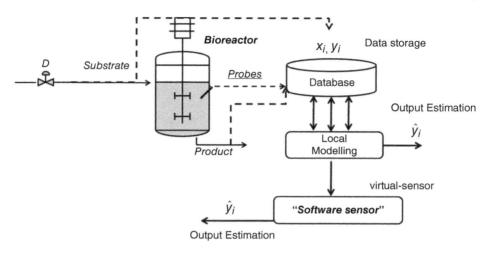

Figure 9.2   Software sensor coupled to the JIT approach in real time.

Table 9.2   Correlation coefficients.

Variable	R^2
S_0	0.95
X_0	0.96
B_0	0.93
P_0	0.98
Q_0	0.94
G_0	0.97
L_0	0.95
Global	0.953

9.5   Results

Numerical simulations were carried out using MATLAB™ library 23 ordinary differential equation (ODE). The parameter values were determined by non-linear regression employing the L–M (Table 9.2). Levenberg–Marquardt (L–M) least squares minimization algorithm, which is a hybrid of the Gauss–Newton and the steepest descent methods. These statistics used absolute values rather than squared differences (as in their originally specified counterparts) [33]. Interpretation of correlation-based measures 0.953 indicates that the model explains 95.3% of the variability in the observed data. With the indices of the agreement, any value (accepting 0.0 and 1.0) is difficult to interpret because of its physical meaning. The following initial conditions for the corresponding concentrations were considered: $S_0 = 0.1\,\text{cm}^2$; $X_0 = 0.15\,\text{cm}^2$; $B_0 = 0.010\,\text{mU mg P}^{-1}$, $P_0 = 0.010\,\text{mU mg P}^{-1}$, $Q_0 = 0.010\,\text{mU mg P}^{-1}$, $G_0 = 0.015\,\text{mU mg P}^{-1}$, $L_0 = 0.0140\,\text{mU mg P}^{-1}$.

The experimental data served to obtain the 36 parameters values of the model developed (see Table 9.1).

The proposed observer provides a good state estimation (Figures 9.3–9.10), the proposed observer gain is $g_1 = 1.5\,\text{mU mg}^{-1}\,\text{h}^{-1}$, and for the extended Luenberger observer is $g_1 = 1.9\,\text{mU mg}^{-1}\,\text{h}^{-1}$. The gain in the two cases is selected in a heuristic way so that the error tends to 0.

For application purposes, the *T. harzianum* concentration was considered as the measured input of the observers. Figures 9.3–9.10 indicate whether the *T. harzianum* concentration is the measurable output, the reconstruction is of six states: *T. harzianum*, *C. cladosporioides*, β-1,3 glucanase, lipases, proteases, and xylanases concentrations, consequently the non-observable variable is chitinases. The following initial conditions for the Luenberger observer and proposed observer are:

$$\hat{x}_0 = [5\ 4\ 0.2\ 0.10\ 0.10\ 0.10\ 0.10]^T \tag{9.24}$$

considering Eqs. (9.8) the structure for the extended Luenberger observer is:

$$\hat{x} = f(\hat{x}) + g(\hat{x}, u) + Q^{-1}(\hat{x})K(y - \hat{y}) \tag{9.25}$$

$$\hat{y} = h(\hat{x}) \tag{9.26}$$

$Q(\hat{x})$ is the observability matrix defined in (3.56)

Figures 9.3–9.9 show the comparison between the actual values and corresponding estimated variables. Based on the presented results, good performance can be observed of the

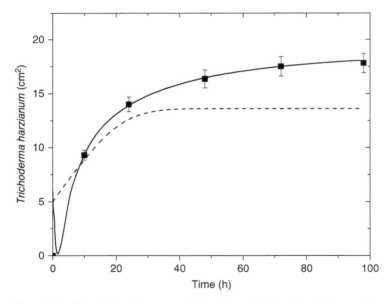

Figure 9.3 Simulation of mycoparasitism. Dynamic behavior of the experimental data of *T. harzianum* concentration (symbols) and the numerical simulations (virtual sensor —, extended Luenberger observer _ _ _).

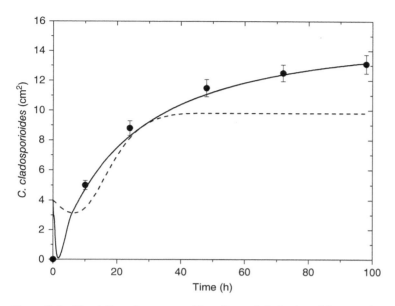

Figure 9.4 Simulation of mycoparasitism. Dynamic behavior of the experimental data of *C. cladosporioides* concentration (symbols) and the numerical simulations (virtual sensor −, extended Luenberger observer _ _ _).

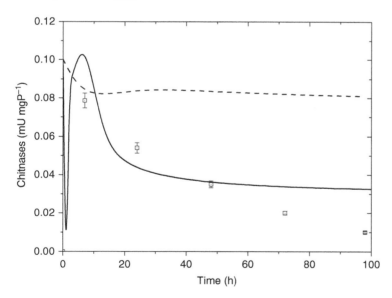

Figure 9.5 Simulation of mycoparasitism. Dynamic behavior of the experimental data of chitinases concentration (symbols) and the numerical simulations (virtual sensor −, extended Luenberger observer _ _ _).

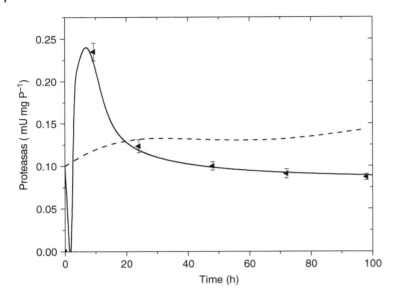

Figure 9.6 Simulation of mycoparasitism. Dynamic behavior of the experimental data of proteases concentration (symbols) and the numerical simulations (virtual sensor −, extended Luenberger observer _ _ _).

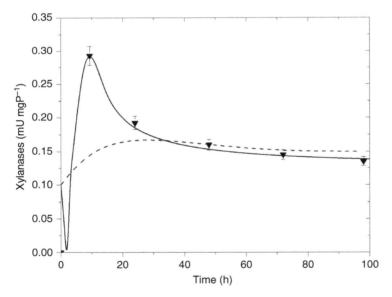

Figure 9.7 Simulation of mycoparasitism. Dynamic behavior of the experimental data of xylanases concentration (symbols) and the numerical simulations (virtual sensor −, extended Luenberger observer _ _ _).

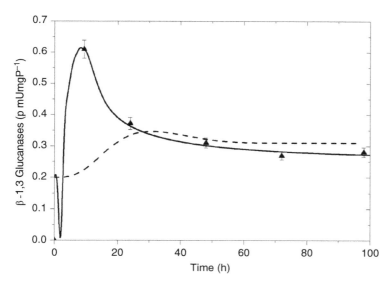

Figure 9.8 Simulation of mycoparasitism. Dynamic behavior of the experimental data of β-1,3 glucanases concentration (symbols) and the numerical simulations (virtual sensor —, extended Luenberger observer _ _ _).

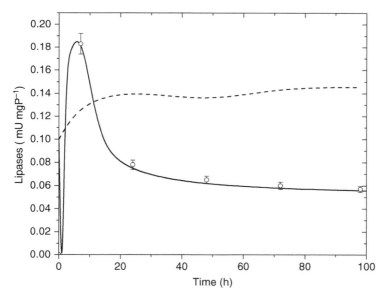

Figure 9.9 Simulation of mycoparasitism. Dynamic behavior of the experimental data of lipases concentration (symbols) and the numerical simulations (virtual sensor — , extended Luenberger observer _ _ _).

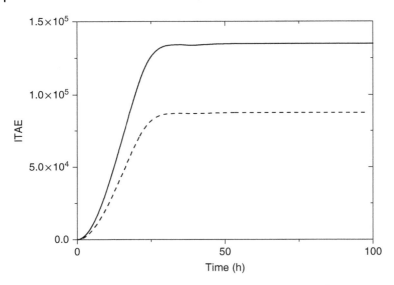

Figure 9.10 The integral time absolute error. Numerical simulations (extended Luenberger observer —, virtual sensor _ _ _).

observer predictor scheme proposed while the only the extended Luenberger observer does not converge. A correct and fast estimation for the reconstruction of *T. harzianum*, *C. cladosporioides*, β-1,3 glucanase, lipases, proteases, and xylanases using the cascade or parallel observer- JIT model is achieved making the proposed system suitable for observer-based control applications.

The trajectories of the proposed methodology converge quickly to the real trajectories, considering modeling errors (a 90.0% of μ_{max} was considered), which are not the case for the extended Luenberger observer. Even with the model (considering modeling errors), the proposed observer produced a stable response, whereas the extended Luenberger observer gave a poor response (Figures 9.3, 9.4, 9.6–9.9). Therefore, the approach presented here may be considered as a valid methodology to design a virtual sensor for this class of systems. As the diagram shows, for the first hours, the proposed observer could accurately estimate the *T. harzianum*, *C. cladosporioides*, β-1,3 glucanase, lipases, proteases, and xylanases concentrations as confirmed by the experimental values.

The integral time absolute error (ITAE) is evaluated for two observers [34]. The information from Figure 9.10 suggests that the proposed observer approach produces better results than the extended Luenberger observer. This is confirmed by the estimation error, which is an order of magnitude lower than those observed for the extended Luenberger observer. In order to obtain a reliable virtual sensor that can be used for the prediction and optimization of large scale control systems, careful design of a virtual sensor calibration stage and very informative experiments are required. A virtual sensor and JIT model representing the dynamic behavior of biochemical species in a cell are required. This methodology can be used to evaluate the behavior of the biochemical reaction network in biocontrol and to determine the properties of the dynamic system of networks such as the existence of an equilibrium point, the uniqueness or the multiplicity of equilibrium points, local or global asymptotic stability of equilibrium points; as well as analyzing control of these networks for design biocontrol [35]. Therefore, the approach presented here may be considered as a valid methodology to design an observer for this class of systems.

In fact, the molecular tools available for the species- and strain-specific monitoring of *Trichoderma*, ranging from fingerprinting tools to exogenous markers and immunological approaches, polymerase chain reaction (PCR) as well as "omics" approach, are expensive and do not allow a real-time quantification of the critical variables in biocontrol. This result indicates which can be used (virtual sensor) to follow the presence and performance of candidates as well as biocontrol strains and metabolites, as an alternative to biocontrol in agriculture [36–40].

9.6   Conclusions

The present study developed and investigated a virtual sensor for online monitoring of biological control of *T. harzianum* against *C. cladosporioides*. A new approach for a dynamic kinetic model for biological control through *T. harzianum* against *C. cladosporioides* was proposed. A mathematical model validated experimentally was used as a virtual plant together with the concentration of *T. harzianum* for the real-time estimation of the concentrations of *C. cladosporioides*, β-1,3 glucanases, lipases, proteases, proteolytic enzymes that according to their degree of induction could participate in the antagonistic effect of *T. harzianum* VSL291 against *C. cladosporioides*. On the other hand, the concentration that cannot be estimated was the chitinase, in spite of errors in the evaluation of the parameters and white noise in the *T. harzianum* measurement (disturbance) using *T. harzianum* concentration as measurable output. Results showed that the proposed estimator allows the problem of real-time monitoring to be tackled.

List of Figures

List of Tables

References

1 Cuervo-Parra, J.A., Sánchez-López, V., Ramirez-Suero, M., and Ramírez-Lepe, M. (2011). Morphological and molecular characterization of *Moniliophthora roreri* causal agent of frosty pod rot of cocoa tree in Tabasco, Mexico. *Plant Pathology* 10 (3): 122–127.

2 Kilaru, A., Bailey, B.A., and Hasenstein, K.H. (2007). *Moniliophthora perniciosa* produces hormones and alters endogenous auxin and salicylic acid in infected cocoa leaves. *FEMS Microbiology* 274: 238–244.

3 Djocgoue, P.F., Boudjeko, T., Nankeu, D.J. et al. (2006). Comparative assessment of the resistance of cocoa (*Theobroma cacao* L.) progenies from SNK10 x SNK413; ICS84 x ICS95 to *Phytophthora megakarya* in Cameroon by measuring size of necrotic lesion along the midrib. *Plant Pathology* 5: 329–335.

4 Riascos, D., Quiroga, I., and Hoyos-Carvajal, L. (2011). Análisis de la sintomatología de la roña en gulupa (*Pasiflora edulis* f. *edulis* SIMS). *Agron* 19 (1): 20–30.

5 Goes, A. (1998). Donças fúngicas. In: *Simposio Brasileiro sobre a Cultura do Maracujazeiro*, 208–216. Jaboticabal, Brasil: Faculdade de Ciencias Agrárias e Veternárias, Universidade Estadual Paulista.

6 Schubert, K., Groenewald, J.Z., Braun, U. et al. (2007). Biodiversity in the *Cladosporium herbarum* complex (*Davidiellaceae, Capnodiales*), with standardisation of methods for *Cladosporium taxonomy* and diagnostics. *Studies in Mycology* 58: 105–156.

7 The UniProt Consortium (2008). The universal protein resource (UniProt). *Nucleic Acids Research* 36 (Database issue): D190–D195. https://doi.org/10.1093/nar/gkm895.

8 Cuervo-Parra, J.A., Ramírez-Suero, M., Sánchez-López, V., and Ramírez-Lepe, M. (2011). Antagonistic effect of *Trichoderma harzianum* VSL291 on phytopathogenic fungi isolated from cocoa (*Theobroma cacao* L.) fruits. *African Journal of Biotechnology* 10 (52): 10657–10663.

9 Agamez, R.E., Barrera, V.J., and Oviedo, L.Z. (2009). Evaluación del antagonismo y multiplicación de *Trichoderma* sp. en sustrato de plátano en medio líquido estático. *Acta Biológica Colombiana* 14: 61–70.

10 Chimmiri, V. (2004). Advances in monitoring and state estimation of bioreactors. *Journal of Scientific and Industrial Research* 63 (6): 491–498.

11 López Pérez, P.A., Neria González, M.I., and Aguilar-Lopez, R. (2016). Concentrations monitoring via software sensor for bioreactors under model parametric uncertainty: application to cadmium removal in an anaerobic process. *Alexandria Engineering Journal* 55 (2): 1893–1902.

12 Stephanopoulos, G. and San, K.Y. (1989). Studies on on-line bioreactor identification. I. Theory. *Biotechnology and Bioengineering* 26: 1176.

13 Kadlec, P., Gabrys, B., and Strandt, S. (2009). Data-driven soft sensors in the process industry. *Computers and Chemical Engineering* 33 (4): 795–814.

14 López-Pérez, P.A., Maya-Yescas, R., Peña-Caballero et al. (2013). Software sensors design for a model of a simultaneous saccharification and fermentation of starch to ethanol. *Fuel* 110: 219–226.

15 Rauh, A., Butt, S.S., and Aschemann, H. (2013). Nonlinear state observers and extended kalman filters for battery systems. *International Journal of Applied Mathematics and Computer Science* 23 (3): 539–556.

16 Luenberger, D.G. (1964). Observing the state of a linear system. *IEEE Transactions on Military Electronics* MIL-8: 74–80.

17 Szekeres, A., Leitgeb, B., Kredics, L. et al. (2006). A novel, image analysis-based method for the evaluation of in vitro antagonism. *Journal of Microbiological Methods* 65 (3): 619–622.

18 Nobe, R., Sakakibara, Y., Fukuda, N. et al. (2003). Purification and characterization of laminaran hydro-lases from *Trichoderma viride. Bioscience, Biotechnology, and Biochemistry* 67: 1349–1357.

19 Rawashdeh, R., Saadoun, I., and Mahasneh, A. (2005). Effect of cultural conditions on xylanase production by Streptomyces sp. (strain Ib 24D) and its potential to utilize tomato pomace. *African Journal of Biotechnology* 4 (3): 251–255.

20 Nawani, N., Nirpjit, S.D., and Jagdeep, K. (1998). A novel thermostable lipase from a thermophilic Bacillus sp.: characterization and esterification studies. *Biotechnology Letters* 20 (10): 997–1000.

21 Kunitz, M. (1946). Spectrophotometric method for the measurement of ribonuclease activity. *The Journal of Biological Chemistry* 164: 563–568.

22 Romero-Cortes, T., López-Pérez, P.A., Ramírez-Lepe, M., and Cuervo-Parra, J.A. (2016). Kinetic modeling of mycoparasitism by *Trichoderma harzianum* against *Cladosporium* *cladosporioides* isolated from cocoa fruits (*Theobroma cacao*). *Chilean Journal of Agricultural Research*, ex Agro-Ciencia, 32 (1): 3432–3445.

23 Solomatine, D.P. and Ostfeld, A. (2008). *Data-Driven Modeling: Some Past Experiences and New Approaches*. IWA Publishing Journal of Hydroinformatics.

24 Beg, Q.K., Kapoor, M., Mahajan, L., and Hoondal, G.S. (2001). Microbial xylanases and their industrial applications: a review. *Applied Microbiology and Biotechnology* 56: 326–338.

25 Lu, F., Lu, M., Lu, Z. et al. (2008). Purification and characterization of xylanase from *Aspergillus ficuum* AF-98. *Bioresource Technology* 99: 5938–5941.

26 Knob, A. and Carmona, E.C. (2010). Purification and characterization of two extracellular xylanases from *Penicillium sclerotiorum*: a novel acidophilic xylanase. *Applied Biochemistry and Biotechnology* 162: 429–443.

27 Saptoro, A. (2014). State of the art in the development of adaptive soft sensors based on just-in-time models. *Procedia Chemistry* 9: 226–234.

28 Aguilar-López, R., Mata-Machuca, J., Martinez-Guerra, R., and López-Pérez, P.A. (2009). Uniformly bounded error estimator for bioprocess with unstructured cell growth models. *Chemical Product and Process Modeling* 4 (5).

29 Ge, Z. and Song, Z. (2010). A comparative study of just-in-time-learning based methods for online soft sensor modeling. *Chemometrics and Intelligent Laboratory Systems* 104: 306–317.

30 Haimi, H., Mulas, M., Corona, F., and Vahala, R. (2013). Data-derived soft-sensors for biological wastewater treatment plants. *Environmental Modelling & Software* 47C: 88–107.

31 Liu, Y., Huang, D., and Li, Y. (2011). Development of interval soft sensors using enhanced just-in-time learning and inductive confidence predictor. *Industrial and Engineering Chemistry Research* 51: 3356–3367.

32 Galvão, R.K.H., Araujo, M.C.U., José, G.E. et al. (2005). A method for calibration and validation subset partitioning. *Talanta* 67: 736–740.

33 Vasanth, K.K. (2006). Linear and non-linear regression analysis for the sorption kinetics of methylene blue onto activated carbon. *Journal of Hazardous Materials* B137: 1538–1544.

34 Isaza, J.A., Sánchez-Torres, J.D., Jiménez-Rodrguez, E., and Botero-Castro, H. (2016). A soft sensor for biomass in a batch process with delayed measurements. In: *XVII Latin American Conference of Automatic Control, IFAC*, 334–339. Medellín, Colombia.

35 Angeli, D., De Leenheer, P.E., and Sontag, D. (2007). A petri net approach to the study of persistence in chemical reaction networks. *Mathematical Biosciences* 210: 598–618.

36 Nygren, K., Dubey, M., Zapparata, A. et al. (2018). The mycoparasitic fungus *Clonostachys rosea* responds with both common and specific gene expression during interspecific interactions with fungal prey. *Evolutionary Applications* 11: 931–949.

37 László, K., Liqiong, C., and Orsolya, K. (2018). Molecular tools for monitoring Trichoderma in agricultural environments. *Frontiers in Microbiology* 9: 1599.

38 Romero-Cortes, T., López-Pérez, P.A., Pérez España, V.H. et al. (2019). Confrontation of trichoderma asperellum vsl80 against aspergillus Niger via the effect of enzymatic production. *Chilean Journal of Agricultural & Animal Sciences* 35 (1): 68–80.

39 Romero-Cortes, T., Pérez España, V.H., López-Pérez, P.A. et al. (2019). Antifungal activity of vanilla juice and vanillin against *Alternaria alternata*. *CyTA – Journal of Food* 17 (1): 375–383.

40 Cuervo-Parra, J.A., Pérez-España, V.H., López Pérez, P.A. et al. (2019). *Scyphophorus acupunctatus* (Coleoptera: Dryophthoridae): a weevil threatening the production of agave in Mexico. *Florida Entomologist* 102 (1): 1–9.

10

Virtual Sensor Design for State Estimation in a Photocatalytic Bioreactor for Hydrogen Production

This chapter focusses on the design of a virtual sensor for a class of continuous bioreactor to estimate the production of hydrogen. The proposed mathematical model is suitable for predicting concentrations of biomass, acetate, cadmium in liquid, sulfate, lactate, carbon dioxide, sulfide, and cadmium sulfide, as well as hydrogen production. The proposed model (kinetic) is used as a benchmark plant and extended for continuous operation to analyze the local properties (stability and observability), considering three sets of measured outputs that produce observable subspaces of different dimensions. In addition, we present the design of a virtual sensor based on a nonlinear observer, which is robust against modeling errors, to estimate the observable states of the bioreactor. The convergence of the proposed methodology is analyzed using the Lyapunov stability theory. The state observer performance is shown and compared versus the extended Luenberger observer, obtaining well-estimated concentrations of the proposed observer

10.1   Introduction

Emergent concerns about the sustainability and impact on the global climate of our current energy sources are an unavoidable consequence of our society's heavy dependence on fossil fuels. Attention to bioenergetics processes has greatly increased during the last decade [1]. Biomass has been used as a means to produce heat and light generated from various industrial activities, and advanced technologies. Considerable emphasis has been placed on converting raw biomass into more useful forms of energy that are easy to use and transport [2]. Alternatively, hydrogen is a clean energy carrier, producing water as its only by-product when it burns. The development of hydrogen biotechnology is advancing rapidly as a fuel of increasing importance, in parallel with the demise of fossil fuels [3].

Biological approaches could contribute to large-scale H_2 production, as various microorganisms can produce H_2 under moderate conditions from readily available, renewable substrates, making biological strategies potentially competitive with a chemical process such as reforming and gasification [3]. In recent years there have been studies with different microorganisms that have the capability to produce H_2 from organic matter (i.e. bacteria *Clostridium* sp., and algae *Chlorella* sp.) or from water oxidizing catalyzed by enzymes such as nitrogenase and hydrogenase (i.e. cyanobacteria *Anabaena* sp.).

Control in Bioprocessing: Modeling, Estimation and the Use of Soft Sensors, First Edition.
Pablo Antonio López Pérez, Ricardo Aguilar López, and Ricardo Femat.
© 2020 John Wiley & Sons Ltd. Published 2020 by John Wiley & Sons Ltd.

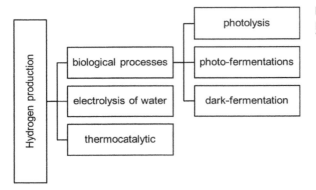

Figure 10.1 The bio-hydrogen production in microorganism "strategies".

Hydrogen is generated by biochemical reactions, mainly in anaerobic–aerobic processes (see, Figure 10.1):

- like biophotolysis of water using green algae and blue–green algae (cyanobacteria) [4, 5]
- direct biophotolysis, and indirect photofermentation, and dark fermentation [6, 7]
- hybrid systems, using dark fermentative and photo-fermentative bioreactors (semicon-ductor photocatalyst) [8, 9]
- bioelectrochemical assisted microbial bioreactors [8, 10]

So far, research has focused on two aspects of obtaining hydrogen sulfide and certain biological catalysts for hydrogen production under different operating conditions [11, 12]. Hydrogen yields by pure or mixed cultures were reported to range from 0.37–2.0 mole-H_2/mole-glucose [13]. Considering the high theoretical yields, several researchers have begun exploring approaches to increase hydrogen production.

Numerous studies have been investigating the possibility of hydrogen production by con-tinuously operated bioreactors [14, 15]. However, it is difficult to develop and, especially, to implement advanced monitoring and control strategies for these bioreactors because of the absence of reliable instrumentation for measuring biological state variables, i.e. substrates, biomass, metabolic enzymes (hydrogenases), and product concentrations. Besides, to opti-mize the reaction and maintain the quality of the product, it would be essential to keep biomass, substrate, and some products at desired values. To this date, PID-type controllers are the most often applied in the process industries. In many cases, the state variables are not measurable on-line (and in real-time) due to the high cost of sensors and extreme oper-ating conditions; these facts, together with the nonlinearity and parameter uncertainty of the bioprocesses, require an enhanced modeling effort, modern state estimation, and iden-tification strategies [16].

Several studies on state estimation and software sensors of microbial growth processes have been reported in the literature [17]. Because of the nonlinear aspects of the biopro-cesses dynamics, observability and controllability analysis are rather complex [18]. There-fore, the theory of observers for nonlinear systems is not yet as fully developed as for linear systems. The development of observability conditions for nonlinear systems is a challeng-ing problem (even for accurately known systems) that has received a considerable amount of attention [19].

Recently, a number of studies have been reported concerning the software sensor design in hydrogen production, for example, a nonlinear observer is proposed and designed, which gives an estimate of both biomass concentration and specific growth rate from measurements of the produced hydrogen volume [19]. Furthermore, kinetic approaches revealed the innovative implementation of an orthogonal least squares algorithm for the calculation of the reaction kinetics involving precise information and the uncertainties of the obtained results; this method was applied to evaluate the reaction rate and, therefore, fractional conversion of methane [20]. In addition, product concentrations of propionate, acetate, and biomass were estimated by an asymptotic online observer from measurements of gas composition in H_2, CO_2, and gas flow rate [21]. In an activated sludge process for the production of bio-hydrogen, the biomass concentration of each group (significantly related to lactic acid production, whereas the evolution of acidogenic microorganisms was related to hydrogen, acetic acid, and butyric acid production) was estimated by modeling the experimental data [22].

In order to solve the problems described, in this chapter, a state observer for a class of continuous bioreactor to estimate the photocatalytic production of hydrogen is proposed to provide monitoring for a class of bioreactor warranting that estimation errors converge whenever the observer gains are adequately chosen in order to compensate the modeling errors. The mathematical model proposed was used as a benchmark plant and extended for continuous operation to analyze the local observability properties. Additionally, we present a nonlinear observer, which is robust against modeling errors, to estimate the observable states of the bioreactor. The convergence of the proposed methodology was analyzed using the Lyapunov stability theory. Finally, numerical experiments were done to show the performance of the proposed observer.

10.2 Material and Methods

10.2.1 Methods

Sulfate concentration was determined by a turbidimetric method based on barium precipitation. The method gives accurate measurements to a few $mg\,l^{-1}$ of sulfate [23]. Lactate and acetate concentrations were evaluated by HPLC in a Shimadzu LC10Ai chromatograph with a UV detector ($\lambda = 210\,nm$) and a BioRad HPLC Organic Acid Analysis Column. Elution was made at a $0.7\,ml\,min^{-1}$ flow rate, using a solution of sulfuric acid/water (0.33/0.67) as a mobile phase. Lactic acid (60% v/v Sigma-Aldrich) and sodium acetate (99.0% Sigma-Aldrich) were used as standards. A 40 ml aliquot of fermented broth from each bottle was centrifuged at 13 000 rpm. The standard deviation of eight measurements was within 9% of the mean value. The bacterial biomass growth and biofilm were evaluated by a dry weight method. A colorimetric assay method was used for sulfide analysis [24]. The supernatant was recovered and filtrated through a nitrocellulose membrane, $0.22\,\mu m$ pore-diameters. The filtrated sample was diluted with 2% (v/v) HNO_3 to determine the cadmium content while the cell pellets were discarded. Cadmium was assayed by atomic absorption spectrophotometry (AAS) (Atomic Absorption Spectrometer Specter AA-20 plus, Varian) using a lamp exclusive for cadmium (Hollow Cathode Lamp Element:

Cadmium, Aurora Instruments). Hydrogen gas was analyzed with a Shimadzu GC-8A gas chromatograph equipped with a thermal conductivity detector and a Hewlett Packard HP.

10.2.2 *Desulfovibrio Alaskensis* 6SR

The microorganism was isolated from a biofilm sited in the inner face of an oil pipeline [25]. The microorganism was in a sterile NaOH solution (0.5 M). Serum bottles (500 ml) were provided with 45 ml of the medium and sterilized (121 °C, 15 minutes). An inoculum of 20% (v/v) of a 36 culture was used for the cadmium removal studies. A stock cadmium acetate solution was prepared to have 10 000 mg l^{-1} of cadmium. The solution was sterilized through membrane filtration (pore size 0.22 μm). The serum bottles with the culture medium were spiked with the stock solution to obtain 140 mg l^{-1} of cadmium. A bottle was taken every 24 hours with cadmium concentration, for analyses of all; sulfate, sulfide (H$_2$S), biomass, lactate, acetate, CO$_2$, H$_2$, and SCd, cadmium concentrations. The experiment consisted of eight bottles (and two duplicates/bottles). The batch cultures were performed in anaerobic conditions at 37 °C for 168 hours (seven days).

10.3 Mathematical Model Development

10.3.1 Basic Concepts

The use of new technology to study sulfide hydrogen splitting by using semiconductor photoelectrodes and photocatalyst for hydrogen production is one of the most significant scientific trends. Both the catalytic activity of the semiconductor and the design of the reactor, influence the possibility of applying photocatalytic reactions to industrial processes. Different types of reactors have been tried but, so far, a successful laboratory photocatalysis setup has not been scaled up to an industrially relevant scale [26]. This chapter presents an alternative to a new configuration of different processes (biological, physical and chemical) for hydrogen production using industrial wastewater by the batch anaerobic process (see, Figure 10.2).

The conceptual model consists of the following:

a) Operation of biological sulfate reduction for cadmium removal (the reduction of sulfate to hydrogen sulfide is brought about by *Desulfovibrio alaskensis* 6SR).
b) The CdS (precipitate) produced is a well-characterized semiconductor which is capable of absorbing visible light [26], for instance, H$_2$S is often used as a sacrificial electron donor for the photocatalytic production of H$_2$. However, based on previous works, another possible mechanism for the photocatalytic production of H$_2$ with visible light and CdS is feasible [27, 28].
c) Finally, a xenon light source was applied to irradiate the aqueous solution of H$_2$S, in which CdS is suspended, which leads to the production of H$_2$.

10.3.2 Proposed Model

In order to improve process understanding or performance, different tools can be considered, as simulators able to reproduce system behavior, software sensors, virtual sensors,

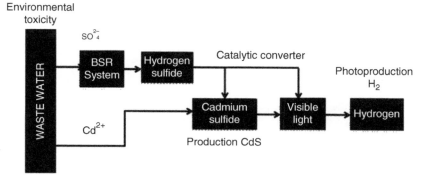

Figure 10.2 Conceptual model.

observer state, which allow an estimation of an unmeasured signal to be obtained to maintain optimal conditions. All these tools rely on a representation of the considered system, which is generally a mathematical model of the process [29]. However, the unstructured kinetic models are frequently employed for modeling simulation of microbial growth, substrate consumption and product formation, Monod, Gompertz, logistic, and Luedeking–Piret models have been used to determine the kinetic parameters for H_2 production [30, 31]. Some investigators developed mathematical models that included the enzyme deactivation during the hydrolysis of the insoluble substrate, these are complex models whose solutions cannot be solved analytically [32].

A mathematical model was developed to describe the kinetics of H_2 production in batch systems (dilution rate $= 0$), this model was used as a benchmark production plant and extended for continuous operation (dilution rate $\neq 0$). In addition, the model includes three processes:

a) The microbial sulfate reduction (the Levenspiel inhibition model was modified to describe sulfate, sulfide, biomass, as well as the acetate, and CO_2 production, with lactate consumption, was obtained by a combination of the Moser–Boulton models and biomass concentration).
b) The mass balance describing the removal of cadmium and precipitation SCd is given by a modified Haldane–Levenspiel model.
c) Production hydrogen by: (i) Monod–Lounge equation in combination with SCd and sulfide concentrations [24], (ii) the other model used was the Gompertz equation (10.5) widely used to describe the changes in hydrogen production [33, 34].

Kinetic terms:

$$v_1 = \left(1 - \frac{P}{k_p}\right)^{\alpha} \left[\frac{S}{k_s + S}\right] X L^{\varepsilon} \tag{10.1}$$

$$v_2 = \left[\frac{k_{ace}}{A + k_{ace}}\right] \left[\frac{L^{\delta}}{k_{lac} + L^{\delta}}\right] X \tag{10.2}$$

$$v_3 = \left(1 - \frac{SCd}{k_{SCd}}\right)^{\gamma} \left[\frac{Cdl}{k_1 + Cdl + \frac{Cdl^2}{k_2}}\right] X P \tag{10.3}$$

Hydrogen production kinetics:

$$v_4 = k_{maxH_2} \left(1 - \frac{P}{k_p}\right)^\beta \left[\frac{SCd^\theta}{k_{SCd} + SCd^\theta}\right] exp\left[-\frac{\vartheta}{Y_{H_2/S}}\right] X L \tag{10.4}$$

$$v_4{}^* = exp\left(-exp\left(\frac{R}{H_2}\right)(\Omega - t) + 1\right) \tag{10.5}$$

Balances:

Sulfate mass balance (S):

$$\frac{dS}{dt} = D(S_{in} - S) - \frac{k_{max\,spx}}{Y}v_1 \tag{10.6}$$

Sulfide mass balance (P):

$$\frac{dP}{dt} = -DP + \frac{k_{max\,spx}}{Y_p}v_1 \tag{10.7}$$

Biomass balance (X):

$$\frac{dX}{dt} = -DX + k_{max\,spx}v_1\,(Cdl^{2+})^{1/\kappa} - k_d\,X\,L^\varepsilon \tag{10.8}$$

Lactate mass balance (L):

$$\frac{dL}{dt} = D(L_{in} - L) - \frac{k_{max\,LA}}{Y_{L/X}}v_2 \tag{10.9}$$

Acetate mass balance (A):

$$\frac{dA}{dt} = -DA + \frac{k_{max\,LA}}{Y_{A/X}}v_2 \tag{10.10}$$

Carbon dioxide mass balance (I):

$$\frac{dI}{dt} = -DI + \frac{k_{max\,LI}}{Y_{I/X}}v_2I \tag{10.11}$$

Cadmium in liquid mass balance (Cdl):

$$\frac{dCdl}{dt} = D(Cdl_{in} - Cdl) - k_{maxCdl}\,v_3 \tag{10.12}$$

Cadmium sulfide (CdS) mass balance:

$$\frac{dCdS}{dt} = -D(CdS) + k_{maxCdS}\,v_3 \tag{10.13}$$

Hydrogen mass balance (H₂):

$$\frac{dH_2}{dt} = -D(H_2) + v_4H_2 \tag{10.14}$$

where $k_{max,j}$ = maximum specific rate for j concentrations, Y = yield constant, k_s = Monod saturation constant for sulphate [$SO_4{}^{2-}$], k_p = inhibition constant for undissociated hydrogen sulphide [H_2S], k_{lac} = inhibition parameter of lactate, k_{ace} = inhibition parameter of lactate, k_{Cd} = inhibition constant for cadmium, $Y_{L/X}$ = yield coefficient: lactate/biomass, $Y_{A/X}$ = yield coefficient: acetate/biomass, α, β and γ = exponential term for Luong model, δ and θ = exponential term for Moser model, η exponential term for cadmium concentration, ε = exponential term for lactate concentration, D = dilution rate, k_d = mortality constant,

Cdl_{in} = initial cadmium concentration, k_1 = saturation coefficient, k_2 = inhibition constant for haldane, ϑ = parameter in function of light intensity, $Y_{\vartheta/H2}$ the hydrogen on light intensity yield, R = maximum production rate, Ω = lag time, e = constant (2.718282).

10.3.3 Determination of Kinetic Parameters

Equations (10.1)–(10.14), were used for the parameter fitting with the following initial concentrations and conditions: S_0 = 5655 mg l^{-1}, P_0 = 29 mg l^{-1}, X_0: = 20 mg l^{-1}, L_0 = 464 mg l^{-1}, A_0 = 0.01 mg l^{-1}, I_0 = 0.01 mg l^{-1}, SCd_0 = 0 mg l^{-1}, Cdl_0 = 140 mg l^{-1}, H_2 = 0 mg l^{-1}. The software Model-Maker® (based on the Levenberg–Marquardt optimization approach) was employed for the nonlinear fitting data [35].

The performance of the proposed mathematical model was evaluated via a dimensionless coefficient of efficiency (Ψ).

$$\Psi = 1 - \frac{\sum_{i=1}^{N}[\Psi(t_i) - O(t_i)]}{\sum_{i=1}^{N}|O(t_i) - \overline{O}|} \tag{10.15}$$

where:

- $\Psi(t_i)$ is the numerically simulated value of a variable at time t_i
- $O(t_i)$ the observed value of the same variable at time t_i
- $\overline{O}$ is the mean value of the observed variable and N is the data number
- Ψ varies between (0, 1], a positive value of Ψ represents an *acceptable* simulation whereas $\Psi > 0.9$ represents a *good* simulation, furthermore, a standard correlation coefficient was also calculated (see, Table 10.1).

Table 10.1 Procedure for the adjustment of kinetic parameters of the proposed model for the removal of cadmium.

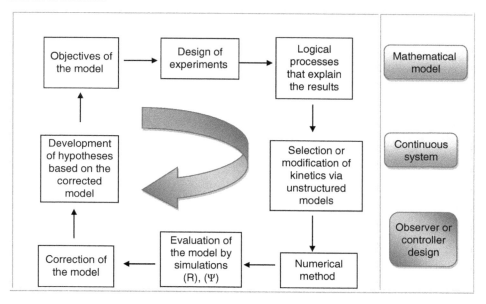

10.4 Virtual Sensor Design

A virtual sensor is a software sensor as opposed to a physical or hardware sensor. The virtual sensors provide indirect measurements (that, by themselves, are not physically measurable) in real conditions by combining sensed data from a group of heterogeneous physical sensors (in real time). Signals from these individual sensors can be used in calculations within a virtual sensor to variables state reconstruction. It is inevitable that some deviation occurs between the trajectories of a real system and the predictions of a mathematical model. Moreover, there are disturbance signals in virtually all engineering applications and the combined effect of these problems generates the need to study the design of virtual sensor techniques that were robust to modeling appearing errors [36]. A schematic representation of this method is shown in Figure 10.3.

Actually, this fact can produce serious problems when some control and monitoring methodologies are implemented. In this case, the problem is to design an asymptotic state estimator of the unknown variables (biomass, lactate, acetate, cadmium, cadmium sulfide, and production hydrogen) based on on-line measurements of the sulfate, biomass, and sulfide.

In Figure 10.3, the output measurements are used to feedback the virtual-sensor with the real process. The algorithm of the virtual-sensor uses these signals as the online available measurement data, noting that the concentration of the sulfate, biomass, and sulfide are considered as the measured system output and this allows the reconstruction of some state variables.

First, it is necessary to consider a generalized space state representation of the system (10.1)–(10.14):

$$\dot{x} = f(x, u) + \Delta\tau \tag{10.16a}$$

$$y = h(x) = Cx \tag{10.16b}$$

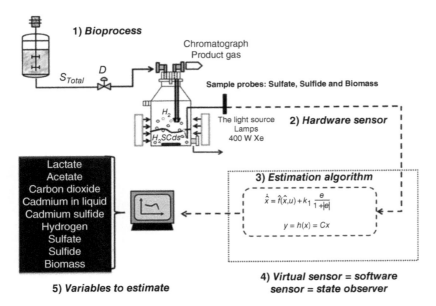

Figure 10.3 Schematic representation of virtual sensor design.

where

$y \in \Re^m$ is the measured output vector, that is the vector of measured states; $x \in \Re^n$ is the vector of state variables; $f(\circ) : \Re^{n+q} \to \Re^n$ is a smooth nonlinear vector function and Lipschitz in x, while $\Delta\tau$ is the additive bounded in the modeling error.

Now, consider the following assumptions, which are common for chemical reactor modeling.

$$f(x) \in C^\infty f(0) = 0$$

and

$$\|f(x)\| \leq F. \ \|\Delta\tau\| \leq T, \forall x \in \Re_+^9, F, T \leq \infty$$

$$f(x, u) = \begin{bmatrix} D(S_{in} - S) - \dfrac{k_{maxspx}}{Y} v_1 \\[2mm] -DP + \dfrac{k_{maxspx}}{Y_p} v_1 \\[2mm] -DX + k_{maxspx} v_1 Cdl^{1/n} - k_d X L^\varepsilon \\[2mm] D(L_{in} - L) - k_{max\,LA} Y_{L/X} v_2 \\[2mm] -DA + k_{max\,LA} Y_{A/X} \ v_2 \\[2mm] -DI + k_I Y_{I/X} \left[\dfrac{k_I}{A + k_I} \right] v_2 \\[2mm] D(Cdl_{in} - Cdl) - k_{maxCdl} v_3 \\[2mm] -CdS + k_{maxCdl} v_3 \\[2mm] -D(H_2) + v_4 H_2 \end{bmatrix} \tag{10.17}$$

$$x = [S, P, X, L, A, I, Cdl, CdS, H_2]^T \in \Re_+^9 \tag{10.18}$$

Proposition 10.1 The following dynamic system is an observer for the system (10.16a) and (10.6b):

$$u = -k_1 \frac{e}{1 + |e|} - k_2 \text{sign}(e) \int_0^t \exp{-(e)^2} d\tau \ \hat{\dot{x}} = \hat{f}(\hat{x}, u) + k_1 \frac{e}{1 + |e|} \tag{10.19}$$

where

$e = x - \hat{x}$ error for the proposed state observer

k_1 and k_2 are the observer parameters

x: state variables

$\hat{x}$: estimate of the state variables x

$k_1 \frac{e}{1+|e|}$: sigmoid function, also called the sigmoidal curve

The general structure of the state observer for the system (10.16) is:

$$\hat{\dot{x}} = \hat{f}(\hat{x}, u, y) \tag{10.20}$$

Usually, it is required that at least $\|\hat{x} - x\| \to 0$, as $t \to \infty$. However, exponential convergence is also required in some cases [36].

Sketch of Proof of Proposition 10.1

Considering Eqs. (10.16) and (10.19), the observation error dynamic is defined as follows:

$$\dot{e} = f(x, u) - \hat{f}(\hat{x}, u) + \Delta\tau + k_1 \frac{e}{1 + |e|} \tag{10.21}$$

Now, taking into account the following assumptions, and according to the corresponding function properties:

A1

$$\left\| k_1 \frac{e}{1 + |e|} \right\| \le 1$$

Taking into account that the above term is a class of sigmoid function.

A2

$$\|\Delta\tau\| \le \Gamma$$

Let us consider the following Lyapunov candidate function;

$$V = e^T \Omega e = \|e\|_\Omega^2, \Omega = \Omega^T > 0 \tag{10.22}$$

The time derivative along the trajectories of (10.22) is:

$$\dot{V} = \dot{e}^T \Omega e + e^T \Omega \dot{e} \tag{10.23}$$

$$\dot{V} = \left(f(x, u) - \hat{f}(\hat{x}, u) + \Delta\tau - k_1 \frac{e}{1 + |e|} \right)^T \Omega e +$$
$$e^T \Omega \left(f(x, u) - \hat{f}(\hat{x}, u) + \Delta\tau - k_1 \frac{e}{1 + |e|} \right) \tag{10.24}$$

$$\dot{V} = 2e^T \Omega (f(x, u) - \hat{f}(\hat{x}, u) + \Delta\tau) - 2e^T \Omega \left(k_1 \frac{e}{1 + |e|} \right) \tag{10.25}$$

(a) The matrix Ω Q can be expressed as $\Omega = RR^T = MMT$, then

$$\|e^T \Omega (f(x, u) - \hat{f}(\hat{x}, u) + \Delta\tau)\| = \|e^T RR^T (f(x, u) - \hat{f}(\hat{x}, u) + \Delta\tau)\| = \|\tilde{e}^T (\tilde{f} + R^T \Delta\tau)\| \tag{10.26}$$

defining $\tilde{e}^T = e^T R$ $eT = eTM$ and $\tilde{f} = R^T (f(x, u) - \hat{f}(\hat{x}, u)) f = fx,u-f(x,u)$

Then,

$$\|\tilde{e}^T\| = (\tilde{e}^T \tilde{e})^{1/2} \tag{10.27}$$

$$\|\tilde{e}^T\| = (e^T RR^T e)^{1/2} \tag{10.28}$$

$$\|\tilde{e}^T\| = (e^T \Omega e)^{1/2} = \|e\|_\Omega \tag{10.29}$$

Similarly as above:

$$\|\tilde{f}\| = \|f\|_\Omega \tag{10.30}$$

Hence,

$$\|e^T \Omega (f(x, u) - \hat{f}(\hat{x}, u) + \Delta\tau)\| = \|\tilde{e}^T (\tilde{f} + \Delta f)\| \le \|\tilde{e}^T\| \|(\tilde{f} + \Delta\tau)\| = \|e\|_\Omega (\|\tilde{f}\| + \Gamma) \tag{10.31}$$

(b) Taking into account (a),

$$\left\| e^T \Omega \left(k_1 \frac{e}{1 + |e|} \right) \right\| \le \|e\|_\Omega \left(k_1 \left\| \frac{e}{1 + |e|} \right\| \right) \tag{10.32}$$

From (a) and (b),

$$\dot{V} \le 2[\|e\|_\Omega(\ell - g_1)\|e\|_\Omega + (\Gamma - g_2)\|e\|_\Omega] \tag{10.33}$$

$$\dot{V} \le 2[(\ell - g_1)\|e\|_\Omega{}^2 + (\Gamma - g_2)\|e\|_\Omega] \tag{10.34}$$

Now, considering $(\ell - g_1) < 0$ and $(\Gamma - g_2) < 0$, by adequate selection of the observer gains, it can be concluded that $\dot{V} \le 0$; therefore, the convergence of the observer is stable.

10.5   Results and Discussion

The volume of hydrogen produced increases with the increase in cadmium sulfide and sulfide concentration initially; these concentrations can be seen in Figure 10.4. The kinetic analysis for the specific growth rate of biomass was conducted in the batch reactors.

Table 10.2 shows the maximum hydrogen production potentials that are at the concentrations of 90 and 225 ml. The hydrogen yields reported in the literature ranged from 1.97 to 0.4 H_2/biomass (see Table 10.2). In addition, according to the results obtained by Senturk and Buyukgungor et al. [44] at the 24th hour, maximum hydrogen production (2489 ml m^{-3} H_2) was observed. Zagrodnik et al. [45] immobilized *Rhodobacters phaeroides* O. U.001 on porous glass plates or glass beads for photo-H_2 production, 220 ml was achieved in a photobioreactor with a 200 ml culture volume from malic acid. The volumetric hydrogen production (VHPR) from anaerobically fermented was reported to be 229 ml $H_2 L^{-1} d^{-1}$ in Torres et al. [46].

Therefore, in this work an output of 200 ml was obtained, with a time of reaction of approximately 170 hours (7 days), it is noteworthy that this hydrogen is obtained by reuse of toxic waste which shows a clear advantage over previous processes. The initial rate (first 24 hours) of lactate consumption decreased with increasing Cd^{2+} concentration as shown in Figure 10.5. Therefore, in this work an output of 200 ml was obtained, with a time of reaction of approximately 170 hours (7 days), it is noteworthy that this hydrogen is obtained by reuse of toxic waste which shows a clear advantage over previous processes (see, Figure 10.6).

Moreover, in our work, Cd^{2+} removal occurred by precipitation of cadmium sulfide the Cd^{2+} removal was nearly complete within 96 hours when the initial Cd^{2+} concentration was below 140 mg l^{-1}. Numerical simulations were done employing the 23 seconds Matlab™ R2009 library to solve ordinary differential equations. The following initial conditions for the corresponding concentrations were considered: $S_0 = 6000$ mg l^{-1}; $P_0 = 15$ mg l^{-1}, $X_0: = 150$ mg l^{-1}, $L_0 = 500$ mg l^{-1}, $A_0 = 5$ mg l^{-1}, $I_0 = 0.5$ mg l^{-1}, $Cdl_0 = 140$ mg l^{-1}, $CdS_0 = 0$ mg l^{-1}, $H_2 = 0$ mg l^{-1} (Table 10.3).

The parametric identification was illustrated by the correlation coefficients above 0.95 between the experimental values and those given by the model (Table 10.4). The regression coefficient using the model Gompertz with a coefficient of 0.73 indicating that does not have a good fit to the experimental data is shown.

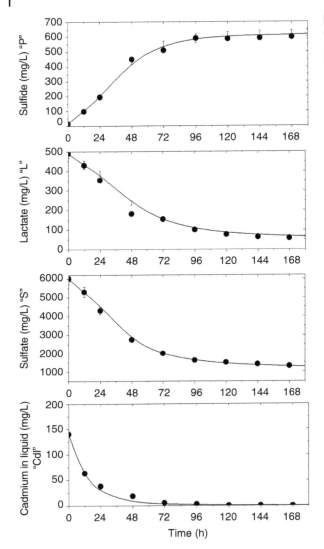

Figure 10.4 Sulfate reduction and cadmium removal, the comparison between batch experimental data (symbols) and model prediction.

The experimental data served to obtain the 28 parameters values of the model developed (see Table 10.3).

The model validation was used by continuous operation, using as a parameter bifurcation dilution rate (D) with the aim of maximizing the production of hydrogen. The maximum hydrogen production by manipulating the dilution rate (D: 0.01 h^{-1}) was 175 ml.

The local stability analysis of the bioreactor was determinate for a dilution rate $D = 0.010\,h^{-1}$, as shown in the Table 10.5, from the above, it can be concluded that the corresponding steady state is locally stable in accordance with the criteria of Lyapunov [47].

Hence, the system (10.1–10.12) is not fully observable but several observable subspaces of different dimensions exist as can be seen for example (Table 10.5):

Table 10.2 Hydrogen productivities reported in literature works.

Author	Substrate	Strain	Reactor type/ conditions maintained	Specific rate of hydrogen	Maximum yield	The volume of hydrogen produced will be maximum
[37]	Centrate of boiling treated sludge	*Consortiums* in boiled treated sludge	Batch reactor	—	1.42	—
[38]	Starch	Sludge from anaerobic digester	Serum bottles	6.54	41.23	—
[39]	Crystalline cellulose	Sludge from anaerobic digester	Serum bottles	1.5	3.2	—
[40]	SnS	Photoelectro-chemical	Coaxial tubular annular	—	—	$90\,ml/\,0.33\,g\,l^{-1}$
[41]	Glucose	Anaerobic	Biofilm reactor 1.41	—	0.4–1.7	$7.61\,(l\,h^{-1})^{-1}$
[42]	Glucose	*C. aceto-butylicum*	Chemostat culture	—	1.97	—
[43]	Glucose	*Clostridium sp*	(CSTR)	1.43	0.85	—
In this work	Lactate	*Desulfovibrio alaskensis*	Batch reactor	0.09	0.9	225 ml

— No reported.

Study Case 10.1 $D = 0.010\,h^{-1}$

- Output: if sulfate concentration is the measurable.
- Reconstruction of six states: sulfate, biomass, sulfide, cadmium in liquid, cadmium sulfide and production hydrogen.
- Unobservable variable: CO_2, lactate, and acetate (Figure 10.7).

Study Case 10.2 $D = 0.010\,h^{-1}$

- Output: if biomass concentration is the measurable.
- Reconstruction states: biomass, sulfide, sulfate, cadmium sulfide and production hydrogen.
- Unobservable variable: lactate, acetate, cadmium in liquid, and CO_2.

For all the considered measured sets, the observability matrix is non-full rank. A nonlinear observer was designed to estimate the observable variables, with the following initial conditions for the observers:

$$\hat{x}_0 = [5755\ 20\ 2\ 540\ 0.01\ 140\ 0.09\ 0.09\ 0.1\,]^T \tag{10.35}$$

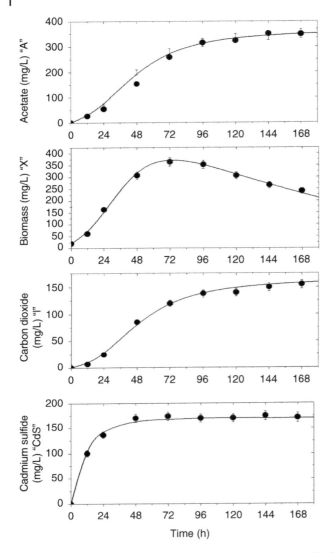

Figure 10.5 Biomass, acetate, cadmium sulfide and carbon dioxide, the comparison between batch experimental data (symbols) and model prediction.

The proposed observer provides a good state estimation (Figure 10.7–10.9), it can be seen that the estimation error of the nonlinear observer is larger than that of the proposed observer, the observer gains are $k_1 = 15\,h^{-1}$ and for nonlinear $k_1 = 15\,h^{-1}$.

The trajectories of the proposed methodology converge quickly to the real trajectories, considering modeling errors (to 95.0% of $k_{max\,spx}$ and 95.0% of k_S were considered), which are not the case for the extended Luenberger observer. Even with the model (considering modeling errors), the proposed observer produced a stable response, whereas the nonlinear observer gave a poor response. Therefore, the approach presented here is a valid methodology to design a virtual sensor for this class of systems [47–54].

Figure 10.6 Hydrogen production, the comparison between batch experimental data (symbols) and model prediction.

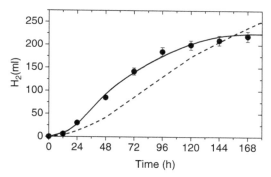

Table 10.3 Kinetic parameters for the mathematical model.

Parameter	Value	Units	Parameter	Value	Units
α	1.76 ± 0.2	Dimensionless	k_p	635.8 ± 100	$mg\,l^{-1}$
γ	2 ± 0.25	Dimensionless	k_s	680 ± 10.0	$mg\,l^{-1}$
β	1.76 ± 0.05	Dimensionless	k_d	0.082 ± 0.02	h^{-1}
δ	2.85 ± 0.03	Dimensionless	k_1	60 ± 2	$1\,h^{-1}\,mg$
ϑ	5.1 ± 0.3	Dimensionless	k_{SCd}	172.59 ± 100	$mg\,l^{-1}$
κ	0.02 ± 0.01	Dimensionless	k_{maxCdl}	1.53	$1\,h^{-1}\,mg$
ε	1.02 ± 0.05	Dimensionless	k_{maxCdS}	1.6	$1\,h^{-1}\,mg$
θ	2.0 ± 0.3	Dimensionless	k_{maxH2}	0.09	$1\,h^{-1}\,mg$
Y	0.13 ± 1.25	Dimensionless	k_2	56.9 ± 2	$mg\,l^{-1}$
$k_{max\,spx}$	0.125 ± 0.05	h^{-1}	$Y_{L/X}$	1.7 ± 0.23	Dimensionless
Y_p	0.9 ± 0.1	Dimensionless	$Y_{A/X}$	1.5 ± 0.23	Dimensionless
k_{ace}	60 ± 5	$mg\,l^{-1}$	$Y_{H2/S}$	0.9 ± 0.03	Dimensionless
k_{lac}	$1.85E4 \pm 500$	$mg\,l^{-1}$	$k_{max\,LA}$	0.05 ± 0.05	h^{-1}
R	120 ± 5	$mg\,l^{-1}$	$k_{max\,LI}$	0.05 ± 0.05	h^{-1}

Table 10.4 Correlation coefficients and model efficiency.

Variable	R^2	Model efficiency(Ψ)
Sulfide	0.94	0.97
Sulfate	0.97	0.97
Biomass	0.94	0.96
Cadmium in liquid	0.96	0.99
SCd	0.95	0.99
Lactate	0.93	0.95
Acetate	0.97	0.98
CO_2	0.97	0.98
H_2	0.95	0.98
Global	0.953	0.974

Table 10.5 Local properties: stability and observability.

Considered equilibrium point: $x_{eq} = [2348, 444, 248, 1903, 0.069, 169, 1470, 2536, 654]^T mg\ L^{-1}$
Eigenvalues: $\lambda = [-0.0141, -23.26, -0.0\ 2, -0.02\ 28 + 0.01i, -0.0\ 2\ 8 - 0.01i, -0.035, -0.01, -0.01, -0.01]^T$
Dilution rate = $0.01h^{-1}$

1) $C = [1\ 0\ 0\ 0\ 0\ 0\ 0\ 0\ 0]^T$	4) $C = [0\ 0\ 0\ 1\ 0\ 0\ 0\ 0\ 0]^T$	7) $C = [0\ 0\ 0\ 0\ 0\ 0\ 1\ 0\ 0]^T$
$rank(O) = 6$	$rank(O) = 4$	$rank(O) = 4$
2) $C = [0\ 1\ 0\ 0\ 0\ 0\ 0\ 0\ 0]^T$	5) $C = [0\ 0\ 0\ 0\ 1\ 0\ 0\ 0\ 0]^T$	8) $C = [0\ 0\ 0\ 0\ 0\ 0\ 0\ 1\ 0]^T$
$rank(O) = 6$	$rank(O) = 3$	$rank(O) = 4$
3) $C = [0\ 0\ 1\ 0\ 0\ 0\ 0\ 0\ 0]^T$	6) $C = [0\ 0\ 0\ 0\ 0\ 1\ 0\ 0\ 0]^T$	9) $C = [0\ 0\ 0\ 0\ 0\ 0\ 0\ 0\ 1]^T$
$rank(O) = 5$	$rank(O) = 4$	$rank(O) = 4$

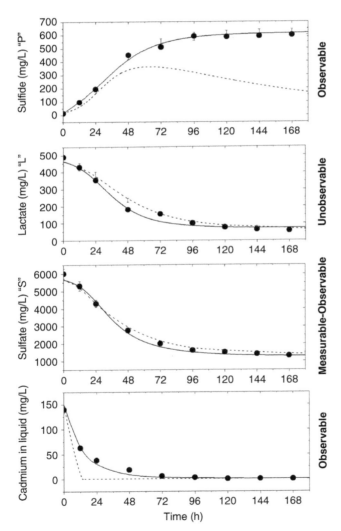

Figure 10.7 Sulfate reduction and cadmium removal, a comparison between batch experimental data (symbols), (– – –) proposed observer and (– – –) extended Luenberger observer.

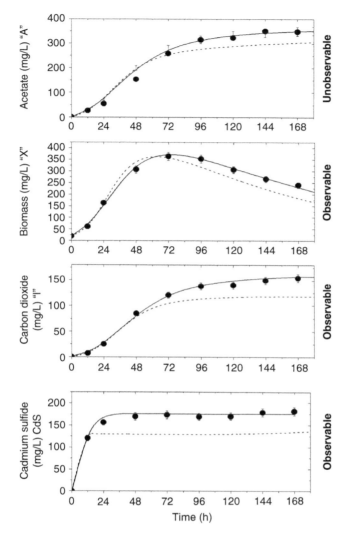

Figure 10.8 Biomass, acetate, cadmium sulfide and carbon dioxide, a comparison between batch experimental data (symbols), (– – –) proposed observer and (– – –) extended Luenberger observer.

Figure 10.9 Hydrogen production, a comparison between batch experimental data (symbols), (– – –) proposed observer and (– – –) extended Luenberger observer.

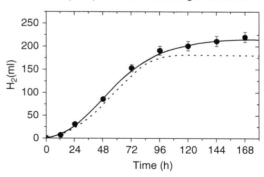

10.6 Conclusions

In this work, we have formulated a dynamic kinetic model for hydrogen production by anaerobic fermentation, using a sulfate-reducing bacterium in a batch stirred bioreactor. The on-line estimation of concentrations of biomass, sulfide, cadmium in liquid, cadmium sulfide, sulfate, lactate, acetate, CO_2, hydrogen production has been investigated, using a virtual sensor based upon a model and considering only on-line measurements of biomass and sulfate concentrations. The positive results, validated with experimental data, display that the estimated concentrations are reliable using a simple model with uncertain parameters. The proposed virtual sensor shows a good performance in comparison with an extended Luenberger observer. The convergence was verified by numerical simulation, with satisfactory performance in accordance with a mathematical stability sketch of the proof. The solution of using a virtual sensor gives promising guidelines to tackle in the future the problem of real-time control of the production of hydrogen.

List of Figures

List of Tables

References

1 Aguilar-López, R., Ruiz-Camacho, B., Neria-González, M.I. et al. (2017). State estimation based on nonlinear observer for hydrogen production in a Photocatalytic anaerobic bioreactor. *International Journal of Chemical Reactor Engineering* 15 (5): 20170004.

2 McKendry, P. (2002). Review paper energy production from biomass (part 1): overview of biomass. *Bioresource Technology* 83: 37–46.

3 Rangel, E., Sansores, E., Vallejo, E. et al. (2016). Study of the interplay between N-graphene defects and small Pd clusters for enhanced hydrogen storage via spillover mechanism. *Physical Chemistry Chemical Physics* 18 (48): 14.

4 Lin, C.N., Wu, S.Y., Lee, K.S. et al. (2007). Integration of fermentative hydrogen process and fuel cell for on-line electricity generation. *International Journal of Hydrogen Energy* 32: 802–808.

5 Kumazawa, S. (2004). Hydrogen production capability in unicellular cyanobacteria. *Plant and Cell Physiology* 45: S23.

6 Tao, Y., Chen, Y., Wu, Y. et al. (2007). High hydrogen yield from a two-step process of dark- and photo-fermentation of sucrose. *International Journal of Hydrogen Energy* 33: 200–206.

7 Sasikala, K., Ramana, C.H.V., and Raghuveer, R.P. (1992). Photoproduction of hydrogen from the wastewater of a distillery by Rhodobacter sphaeroides O.U.001. *International Journal of Hydrogen Energy* 17: 23–27.

8 Akkerman, I., Janssen, M., Rocha, J., and Wijffels, R.H. (2002). Photobiological hydrogen production: photochemical efficiency and bioreactor design. *International Journal of Hydrogen Energy* 27: 1195–1208.

9 Ditzig, J., Liu, H., and Logan, B. (2007). Production of hydrogen from domestic wastewater using a bioelectrochemically assisted microbial reactor (BEAMR). *International Journal of Hydrogen Energy* 32: 2296–2304.

10 Ozmihci, S. and Kargi, F. (2010). Bio-hydrogen production by photo-fermentation of dark fermentation effluent with intermittent feeding and effluent removal. *International Journal of Hydrogen Energy* 35: 6674–6680.

11 Velázquez-Sánchez, H.I., López-Pérez, P.A., Neria-González, M.I., and Aguilar-López, R. (2017). *Enhancement of Bio-Hydrogen Production Technologies by Sulphate-Reducing Bacteria, in Hydrogen Production Technologies* (eds. M. Sankir and N.D. Sankir). Hoboken, NJ, USA: Wiley.

12 Sabate, J., March, S.C., Simarro, R., and Gimenez, J. (1990). Photocatalytic production of hydrogen from sulfide and sulfite waste streams: a kinetic model for reactions occurring in illuminated suspensions of CdS. *Chemical Engineering Science* 45 (10): 3089–3096.

13 Sen, U., Shakdwipee, M., and Banerjee, R. (2008). Status of biological hydrogen production. *Journal of Scientific and Industrial Research* 67: 980–993.

14 Jung, K.W., Kim, D.H., Kim, S.H., and Shin, H.S. (2011). Bioreactor design for continuous dark fermentative hydrogen production. *Bioresource Technology* 102 (18): 8612–8620.

15 Obeid, J., Magnin, J.P., Flaus, J.M. et al. (2009). Modeling of hydrogen production in batch cultures of the photosynthetic bacterium Rhodobacter capsulatus. *International Journal of Hydrogen Energy* 34: 180–185.

16 López-Pérez, P.A., Peña-Caballero, V., Ruiz Camacho, B., and Aguilar-López, R. (2017). Increasing of lipid productivity in microalgae cultures via dynamic analysis and closed-loop operation. *European Chemical Bulletin* 6 (4): 145–150.

17 Rodriguez-Fernandez, M., Mendes, P., and Bang, J.R. (2006). Hybrid approach for efficient and robust parameter estimation in biochemical pathways. *Biosystems* 83: 248–265.

18 García, V.M., López, E., Serra, M. et al. (2010). Dynamic modeling and controllability analysis of an ethanol reformer for fuel cell application. *International Journal of Hydrogen Energy* 35: 9768–9775.

19 Nuñez, S., Garelli, F., and De Battista, H. (2014). Second-order nonlinear observer for biomass concentration and growth rate estimation in batch photo-bioreactors. *International Journal of Hydrogen Energy* 3: 8772–8779.

20 Sciazko, A., Komatsu, Y., Brus, G. et al. (2014). A novel approach to improve the mathematical modeling of the internal reforming process for solid oxide fuel cells using the orthogonal least squares method. *International Journal of Hydrogen Energy* 39: 16372–16389.

21 Aceves-Lara, C.A., Latrille, E., and Steyer, J.P. (2010). Optimal control of hydrogen production in a continuous anaerobic fermentation bioreactor. *International Journal of Hydrogen Energy* 35: 10710–10718.

22 Nuñez, S., Garelli, F., and De Battista, H. (2012). Nonlinear observer for biomass estimation in a biohydrogen production process. *International Journal of Hydrogen Energy* 37 (13): 10089–10094.

23 Cord-Ruwisch, R.A. (1985). Quick method for determination of dissolved and precipitated sulfides in cultures of sulfate-reducing bacteria. *Journal of Microbiological Methods* 4: 33–36.

24 APHA (1975). American public health association, standard methods for the examination of water and wastewater. In: *American Water Works Association Water Pollution Control Federation*, Washington, vol. 1, 800–869.

25 Neria-González, I., Wang, E.T., Ramirez, F. et al. (2006). Characterization of bacterial community associated to biofilms of corroded oil pipelines from the southeast of Mexico. *Anaerobe* 12: 122–133.

26 Arora, M.K., Sahu, N., Upadhyay, S.N., and Sinha, A.S.K. (1999). Activity of cadmium sulfide photocatalysts for hydrogen production from water: role of support. *Industrial and Engineering Chemistry* 38 (7): 2659–2665.

27 Corredor-Rojas, L.M. (2011). Review of Photocatalytic materials for hydrogen production from H2S1. *Ingeniería y Universidad* 15 (1): 171–195.

28 Ryu, S.Y., Balcerski, W., Lee, T.K., and Hoffmann, M.R. (2007). Photocatalytic production of hydrogen from water with visible light using hybrid catalysts of CdS. *Journal of Physical Chemistry C* 111 (49): 18195–18203.

29 Dochain, D. (2003). State and parameter estimation in chemical and biochemical processes: a tutorial. *Journal of Process Control* 13: 801–818.

30 Koku, H., Eroglu, I., Gunduz, U. et al. (2003). Kinetics of biological hydrogen production by the photosynthetic bacterium Rhodobacter sphaeroides O.U. 001. *International Journal of Hydrogen Energy* 28: 381–388.

31 Luedeking, R. and Piret, E.L. (1959). A kinetic study of the lactic acid fermentation. *Journal of Microbial and Biochemical Technology* 1: 393–412.

32 Dahl, C., Rákhely, G., Pott-Sperling, A.S. et al. (1999). Genes involved in hydrogen and sulfur metabolism in phototrophic sulfur bacteria. *FEMS Microbiology Letters* 180 (2): 317–324.

33 Mu, Y., Wang, G., and Yu, H. (2006). Kinetic modeling of batch hydrogen production process by mixed anaerobic cultures. *Bioresource Technology* 97 (11): 1302–1307.

34 Chen, W.H., Chen, S.Y., Khanala, S.K., and Sunga, S. (2006). Kinetic study of biological hydrogen production by anaerobic fermentation. *International Journal of Hydrogen Energy* 3: 2170–2178.

35 Ahearn, T.S., Staff, R.T., Redpath, T.W., and Semple, S.I.K. (2005). The use of the Levenberg–Marquardt curve-fitting algorithm in pharmacokinetic modeling of DCE-MRI data. *Physics in Medicine & Biology* 50: 85.

36 Paulsson, D., Gustavsson, R., and Mandenius, C.F. (2014). A soft sensor for bioprocess control based on sequential filtering of metabolic heat signals. *Sensors* 10 (14): 17864–17882.

37 Cheng, S.S., Bai, M.D., Chang, S.M., et al. (2000) *Studies on the feasibility of hydrogen production hydrolyzed sludge by anaerobic microorganisms*. The Twenty-fifth Wastewater Technology Conference, Yunlin, Taiwan (in Chinese).

38 Lay, J.J., Lee, Y.J., and Noike, T. (2000). Modeling and optimization of anaerobic digested sludge converting starch to hydrogen. *Biotechnology and Bioengineering* 68: 269–278.

39 Lay, J.J., Lee, Y.J., and Noike, T. (1999). Feasibility of biological hydrogen production from organic fraction of municipal solid waste. *Water Research* 33: 2579–2586.

40 Arya, R.K. (2012). Photoelectrochemical hydrogen production using visible light. *International Journal of Energy Research* 2: 2.

41 Zhang, Z.P., Show, K.Y., Tay, J.H. et al. (2008). Biohydrogen production with anaerobic fluidized bed reactors-a comparison of biofilm-based and granule-based systems. *International Journal of Hydrogen Energy* 33: 1559–1564.

42 Saint-Amans, S., Girbal, L., Andrade, J. et al. (2001). Regulation of carbon and electron flow in clostridium butyricum VPI 3266 grown on glucose–glycerol mixtures. *Journal of Bacteriology* 183: 1748–1754.

43 Mizuno, O., Dinsdale, R., Hawkes, F.R. et al. (2000). Enhancement of hydrogen production from glucose by nitrogen gas sparging. *Bioresource Technology* 73: 59–65.

44 Senturk, I. and Buyukgungor, H. (2017). Biohydrogen production by anaerobic fermentation of sewage sludge - effect of initial pH. *Environment and Ecology Research* 5 (2): 107–111.

45 Zagrodnik, R., Seifert, K., Stodolny, M., and Laniecki, M. (2015). Continuous photo-fermentative production by immobilized Rhodobacter sphaeroides OU001. *International Journal of Hydrogen Energy* 40: 5062–5073.

46 Torres, D.G.B., Shaiane Dal', M.L., Andreani, C. et al. (2017). Hydrogen production and performance of anaerobic fixed-bed reactors using three support arrangements from cassava starch wastewater. *Journal of the Brazilian Association of Agricultural Engineering* 37: 160–172.

47 Kaidi, F., Rihani, R., Ounnar, A. et al. (2012). Photobioreactor design for hydrogen production. *Procedia Engineering* 33: 492–498.

48 Skjånes, K., Andersen, U., Heidorn, T., and Borgvang, S.A. (2016). Design and construction of a photobioreactor for hydrogen production, including status in the field. *Journal of Applied Phycology* 28: 2205–2223.

49 Jean-Pierre, M. and Deseure, J. (2019). Hydrogen generation in a pressurized photobioreactor: unexpected enhancement of biohydrogen production by the phototrophic bacterium Rhodobacter capsulatus. *Applied Energy* 239 (C): 635–643.

50 Oey, M., Sawyer, A.L., Ross, I.L. et al. (2016). Challenges and opportunities for hydrogen production from microalgae. *Plant Biotechnology Journal* 14: 1487–1499.

51 Startsev, A.N. (2017). Hydrogen sulfide as a source of hydrogen production. *Russian Chemical Bulletin* 66: 1378–1397.

52 Priatharsini, N.S. (2016). Developments in bio-hydrogen production from algae: a review. *Research Journal of Applied Sciences, Engineering and Technology* 9: 968–982.

53 Reverberi, A.P., Klemeš, J.J., Varbanov, P.S. et al. (2016). A review on hydrogen production from hydrogen sulphide by chemical and photochemical methods. *Journal of Cleaner Production* 136 (B): 72–80.

54 De Crisci, A.G., Moniri, A., and Xu, Y. (2019). Hydrogen from hydrogen sulfide: towards a more sustainable hydrogen economy. *International Journal of Hydrogen Energy* 44 (3): 1299–1327.

Index

a

Acetate   3, 7, 35, 174, 255, 257–259, 267
Asymptotic   25, 85, 87, 93, 121, 129, 130, 158, 173, 228, 257
Asymptotically   79, 84, 85, 87, 88, 91, 93, 142
Adaptive observers   31, 35, 117, 126, 135, 136, 156, 158, 160, 242
Anaerobic   xii, 21, 33, 45, 49, 160, 225, 256, 258, 267, 272
Aerobic   31, 33, 35, 44
Atomic Absorption Spectrometer   257
Automatic   xi, 5, 18, 31, 109, 135, 160, 242
Autonomous systems   73, 79, 85, 87, 93, 98

b

Barbalat's   vii, 94
Batch   6, 17–19, 23, 27, 40, 42, 49, 160, 185, 191, 200, 207, 258
Bifurcation   30, 72, 84, 94, 96, 100, 209, 266
Biochemical   3, 13, 23, 26, 32, 40, 160, 217, 248
Biodiesel   205
Bioethanol   35, 186
Biological control   233, 235, 239
Biomass   5, 15, 17, 20, 27, 33, 35, 42, 46, 50, 131, 160, 171, 185, 206, 213, 257, 262
Bioprocess control   3, 13, 22, 36, 138
Bioreactor   7, 16, 17, 24, 26, 31, 34, 40, 42, 47, 94, 138, 160, 171, 185, 206, 217, 226, 256
Biotechnological   22, 25, 34, 110, 219, 236
Black-box   45, 47

c

Cadmium sulfide   xii, 255, 259, 260, 262, 265, 267, 271, 272
Carboxymethyl-cellulose   186, 188
Cauchy–Schwartz   147, 224
Cell culture   18, 19, 34
Cellulomonas cellulans   186, 202
Chaos   8, 30, 63, 72, 94, 97
Chemostat   25, 185, 186, 190, 191, 201
Chlorella sp.   255
Cladosporium cladosporioides   233, 235–237, 239, 244, 245, 248, 249
Closed loop   140, 145, 149, 154, 158, 171, 172, 176, 177, 181, 225
Complex eigenvalues   71
Controller signal   28
Condition number   205, 206, 211–213
Continuous   xii, 24, 27, 31–33, 36, 40, 41, 73–81, 86, 87, 91, 93, 94, 96, 97, 100, 113, 116–119, 122–124, 127, 137, 139, 140, 150–152, 155, 160, 171, 181, 185, 189, 190, 200, 201, 205–207, 210, 211, 213, 218, 220, 226, 234, 255, 257, 259, 261, 266
Controllability   xi, 111, 135, 138–141, 146, 256
Controllable   34, 138–140, 143, 145–147, 159, 173
Convergence   25, 30, 31, 38, 84, 86, 93, 94, 99, 110, 118, 119, 126, 129, 151, 158, 222, 233, 240, 241, 255, 257, 272
Correlation coefficients   177, 226, 243, 265, 269

Control in Bioprocessing: Modeling, Estimation and the Use of Soft Sensors, First Edition.
Pablo Antonio López Pérez, Ricardo Aguilar López, and Ricardo Femat.
© 2020 John Wiley & Sons Ltd. Published 2020 by John Wiley & Sons Ltd.